Endotoxemia and Endotoxin Shock: Disease, Diagnosis and Therapy

Contributions to Nephrology

Vol. 167

Series Editor

Claudio Ronco Vicenza

Endotoxemia and Endotoxin Shock

Disease, Diagnosis and Therapy

Volume Editors

Claudio Ronco Vicenza

Pasquale Piccinni Vicenza

Mitchell H. Rosner Charlottesville, Va.

22 figures, 4 in color, and 10 tables, 2010

Basel · Freiburg · Paris · London · New York · Bangalore · Bangkok · Shanghai · Singapore · Tokyo · Sydney

Claudio Ronco
Department of Nephrology
Dialysis & Transplantation
International Renal Research Institute
San Bortolo Hospital
IT-36100 Vicenza (Italy)

Pasquale Piccinni
Department of Nephrology
Dialysis & Transplantation
International Renal Research Institute
St. Bortolo Hospital
IT-36100 Vicenza (Italy)

Mitchell H. Rosner
Division of Nephrology
University of Virginia Health System
Charlottesville, VA 22908

Library of Congress Cataloging-in-Publication Data

Endotoxemia and endotoxin shock : disease, diagnosis, and therapy / volume editors, Claudio Ronco, Pasquale Piccinni, Mitchell H. Rosner.
p. ; cm. -- (Contributions to nephrology, ISSN 0302-5144 ; v. 167)
Includes bibliographical references and indexes.
ISBN 978-3-8055-9484-4 (hard cover : alk. paper) -- ISBN 978-3-8055-9485-1 (e-ISBN)
1. Endotoxemia. 2. Septic shock. I. Ronco, C. (Claudio), 1951- II. Piccinni, Pasquale. III. Rosner, Mitchell H. IV. Series: Contributions to nephrology, v. 167. 0302-5144 ;
[DNLM: 1. Endotoxemia--therapy. 2. Endotoxemia--diagnosis. 3. Hemoperfusion--methods. 4. Polymyxin B--therapeutic use. 5. Shock, Septic--diagnosis. 6. Shock, Septic--therapy. W1 CO778UN v.167 2010 / WC 240 E557 2010]
RC182.S4E527 2010
616.9'4407--dc22
2010017258

Bibliographic Indices. This publication is listed in bibliographic services, including Current Contents® and Index Medicus.

www.karger.com
Printed in Switzerland on acid-free and non-aging paper (ISO 9706) by Reinhardt Druck, Basel
ISSN 0302–5144
ISBN 978–3–8055–9484–4
e-ISBN 978–3–8055–9485–1

Contents

Retraction
"Acute Heart Failure Treatment: Traditional and New Drugs" by Gheorghiade M, Palazzuoli A, Ronco C. Contrib Nephrol, 2010;165;112-128.
This chapter of a previous volume of Contributions to Nephrology has been retracted at the authors' request. A miscommunication between the corresponding author and the co-authors resulted in the publishing of an unfinished article.

Preface

Several signs and symptoms in sepsis are due to the presence of endotoxin in the circulation. Both in animal and human models there is an evident immunological response to the bacterial invasion of the host and the consequent release of endotoxin into the bloodstream. The presence of endotoxin in the circulation leads to altered cardiovascular function, lung dysfunction and acute kidney injury, often characterizing a clinical picture of sepsis and septic shock. This humoral nature of the syndrome makes it logical to try to remove the circulating endotoxin as much as possible in order to mitigate its biological and clinical effects at the cellular, tissue and organ levels. This can be achieved today with a very specific hemoperfusion process utilizing cartridges with immobilized polymyxin B in an extracorporeal circuit. This approach seems to provide for a significant removal of endotoxin with a significant reduction of its circulating levels.

The basic mechanisms, rationale and the clinical results of this new therapeutic approach are summarized in the present volume. The contributors of this book represent a group of outstanding investigators whose studies have helped expand the scientific knowledge about this field. The clinical effects reported in several chapters demonstrate a mitigation of the septic cascade in the early phases, with amelioration of the prognosis and outcome in septic patients treated with this specific form of hemoperfusion. Recent clinical trials seem to confirm the expectations showing a reduction of mortality in patients with early signs of abdominal sepsis due to recent surgery. This opens new avenues for specific interventions in sepsis and, once more, represents important material for a book in the *Contributions to Nephrology* series.

We would like to thank the authors and all the contributors for the enormous effort and the quality of their scientific chapters. We also would like to thank all who made this publication possible and especially Karger for the outstanding editorial assistance.

We feel this book will be a milestone in the field of extracorporeal therapies in sepsis and will be a companion for both basic scientists and clinical professionals for their continuous educational improvement.

Claudio Ronco, Vicenza
Pasquale Piccinni, Vicenza
Mitchell H. Rosner, Charlottesville, Va.

Ronco C, Piccinni P, Rosner MH (eds): Endotoxemia and Endotoxin Shock: Disease, Diagnosis and Therapy. Contrib Nephrol. Basel, Karger, 2010, vol 167, pp 1–13

Endotoxin in the Pathogenesis of Sepsis

John C. Marshall

Department of Surgery, University of Toronto, and the Li Ka Shing Knowledge Institute, St. Michael's Hospital, Toronto, Ont., Canada

Abstract

The word 'sepsis' is a descriptive term that denotes the clinical syndrome resulting from the activation of an innate host response to infection. Sepsis is a useful concept that underlines the fact that the morbidity of serious infection arises through the response of the host, rather than through intrinsic cytopathic effects of the microorganism. However, it has proven inadequate as a means to delineate a population of patients who might benefit from therapies that modulate this response. The syndrome is variable in its clinical expression, and not specific for infection as a cause. Emerging insights into the biology of the innate host immune response reveal that the cellular response can be evoked by a variety of stimuli – including both microbial products and host-derived molecules that are normally intracellular – that signal danger to the host. The disconnect between concept and disease that has hampered the conduct of clinical trials is nicely exemplified in the host response to endotoxin. Endotoxemia occurs in many patients with sepsis, but also in many clinical settings that are noninfectious in nature. Moreover, the biologic behavior of endotoxin resembles that of a hormone more than that of a toxin, suggesting that low level endotoxemia may, under some circumstances, be beneficial. Future studies of antiendotoxin strategies in acute illness are more likely to succeed if they recruit patients with endotoxemia, and titrate therapy to an optimal level.

Sepsis is defined as the systemic host response to invasive infection [1]. This seemingly simple definition belies a much more complex biologic reality, for host-microbial interactions are evolutionarily ancient, intimate and fundamentally symbiotic rather than pathologic. To the clinician, however, sepsis lacks this nuance: it is a potentially devastating clinical disorder that poses enormous therapeutic challenges. Within the developed world, sepsis is the leading cause of death for patients admitted to an intensive care unit (ICU), affecting close to one million North Americans annually, and is responsible for the deaths of

Fig. 1. Hippocrates (460–370 BCE), the Greek physician, first used the word sepsis.

more than 200,000 of these patients [2]. Viewed in a global context, sepsis is the process that underlies the leading causes of death in the developing world, e.g. malaria, pneumonia, parasitic diseases, tuberculosis and infantile diarrhea. Therefore, it can legitimately be seen as the leading cause of preventable morbidity and mortality in the world today.

The clinical syndrome of sepsis embodies a large number of overlapping infectious triggers and host responses, and while consensus definitions emphasize the role of infection with viable microorganisms as the sine qua non for the diagnosis, an identical biologic response with identical clinical sequelae can be triggered by noninfectious causes. Indeed an evolving understanding of the intimate and symbiotic interactions of the eukaryotic host and the prokaryotic microbial world render conventional concepts of infection increasingly inadequate.

This brief review addresses the evolution of the concept of sepsis, contemporary understanding about host-microbial interactions and the biologic processes that are responsible for the clinical syndrome, and the specific role that endotoxin plays as a prototypical trigger.

Sepsis: A Conceptual History

The word 'sepsis' is of Greek origin and is originally attributed to Hippocrates (460–370 BCE) (fig. 1) [3]. Hippocrates held that living organisms die in one of

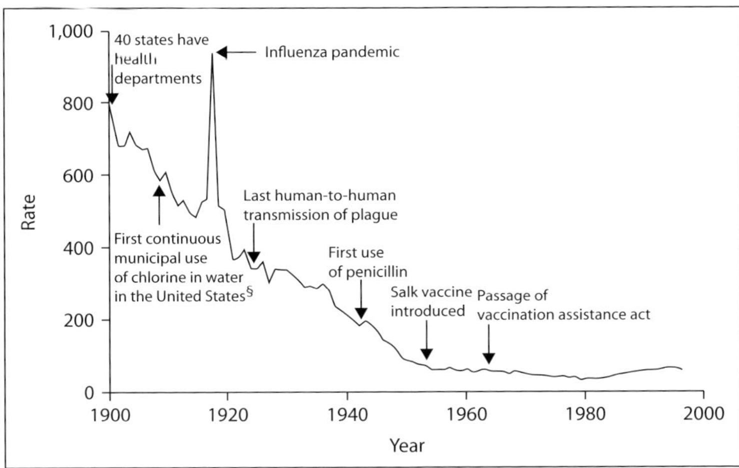

Fig. 2. Incidence of death from infection over the 20th century. Rates dropped precipitously during the first half of the century, primarily as a result of advances in public health measures. The advent of antibiotics and intensive care units had a much more modest impact on mortality at the population level. [6].

two fundamental ways. Sepsis was the process of death that produced ill health, and was exemplified by putrefaction, rot and a foul smell. Pepsis, on the other hand, was tissue breakdown that was life-giving and occurred when food was digested or when grapes were fermented to produce wine. His ideas antedated the articulation of the germ theory of disease by more than two thousand years, and so spoke to the consequences of tissue breakdown rather than its causes.

Work in the 19th century by Semmelweiss, Pasteur and Lister established that infection could be transmitted from one patient to another, and arose through the proliferation of microscopic organisms within the host. Conversely, their studies showed that transmission of infection could be prevented by adherence to principles of asepsis. Articulation of the germ theory of disease led to profound changes in patterns of human illness (fig. 2). Infection could be prevented through measures such as sterilization or pasteurization of milk, or through vaccination with killed or attenuated organisms; it could be successfully treated using an expanding repertoire of antimicrobial agents that could selectively kill the invading organism. Because the most dramatic and important examples of the earlier concept of sepsis were scourges such as the plague, smallpox and a host of other infectious diseases, it was a logical conceptual step to consider sepsis as the clinical manifestation of infection, and the two words came to be used synonymously. As recently as the 1970s, medical dictionaries defined sepsis as 'the presence of pus-forming organisms in the bloodstream'.

Fig. 3. Richard Pfeiffer (standing), the German microbiologist credited with the discovery of endotoxin picture here with Robert Koch (seated).

However, within decades of the identification of bacteria as the transmissible agents of infectious disease, it began to become apparent that the clinical sequelae of infection resulted from processes more complicated than the simple uncontrolled proliferation of microorganisms within the host. The German microbiologist, Richard Pfeiffer (fig. 3), during studies of the pathogen *Vibrio cholerae*, found that even killed vibrios could evoke illness in laboratory animals. He hypothesized that the culprit was a toxin in the bacterial cell wall. Because it was endogenous to the microorganism and toxic to the mammalian host, he termed this factor 'endotoxin'.

The development of antibiotics in the early twentieth century provided further support for the notion that sepsis was a bacterial phenomenon that could be cured by killing the microorganism. Several emerging lines of evidence suggested that things were not so simple.

First, it is apparent that even in the pre-antibiotic era, the majority of patients with serious infections, such as pneumonia, survived their illness [4]. Conversely, the widespread introduction of antibiotics into hospitals did not reduce either the rates or mortality of infection, but simply altered the predominant infecting organisms [5]. Population data on the lethality of infection over the 20th century confirm that the largest reduction in mortality occurred in the first half of the century, prior to the introduction of antibiotics or ICUs, and was more credibly linked to the widespread adoption of improved public health measures [6] (fig. 2).

Second, advances in understanding the biology of the response to infection revealed that many of the cardinal features of infection, e.g. fever, leukocytosis and the characteristic hemodynamic derangements, are mediated not through the direct effects of toxins from the microorganism, but rather indirectly through the activity of factors synthesized and released by the host [7, 8]. Clinical studies showed that the clinical features of infection with Gram-positive and Gram-negative organisms were indistinguishable [9] and similar to those evoked by isolated viral infection [10], or even the infusion of sterile stress hormones [11] or isolated proinflammatory cytokines [12] into healthy volunteers.

Finally, epidemiologic studies revealed that prognosis in critically ill patients is influenced more by the intensity of the septic response than by factors related to the site or bacteriology of invasive infection [13]. We addressed this question in a cohort study of 211 critically ill patients who remained in an ICU for at least 2 days. The presence of infection was diagnosed using exclusively microbiologic criteria that identified the presence of a microorganism in normally sterile tissues, without reference to the response evoked in the host. The magnitude of the clinical response was quantified using a sepsis score that measured increasing severity in five separate domains: temperature, white cell count, alterations in consciousness, increased cardiac output and insulin resistance. While both the development of infection and the expression of a systemic septic response (measured using the sepsis score) correlated with an increased risk of ICU mortality, in patients with documented infection (fig. 4) and in those with a significant septic response (table 1), only the severity of the response predicted ultimate ICU mortality.

Contemporary terminology reflects this awareness that the stimulus – infection – and the response it evokes must be differentiated. Thus, infection is defined as the invasion of normally sterile tissues by microorganisms. The response to that event may be a local response or a disseminated systemic response. In the latter case it is termed 'sepsis'; therefore, sepsis is the systemic host inflammatory response to infection. This response is an adaptive one that can aid the host in clearing the infection; however, it can also have maladaptive consequences. 'Severe sepsis' defines the development of organ dysfunction in association with a septic response, while 'septic shock' denotes accompanying cardiovascular derangements that impair tissue perfusion. Finally, since the response is not necessarily specific for infection, the concept of the Systemic Inflammatory Response Syndrome (SIRS) was articulated to describe the clinical syndrome independent of its cause [1].

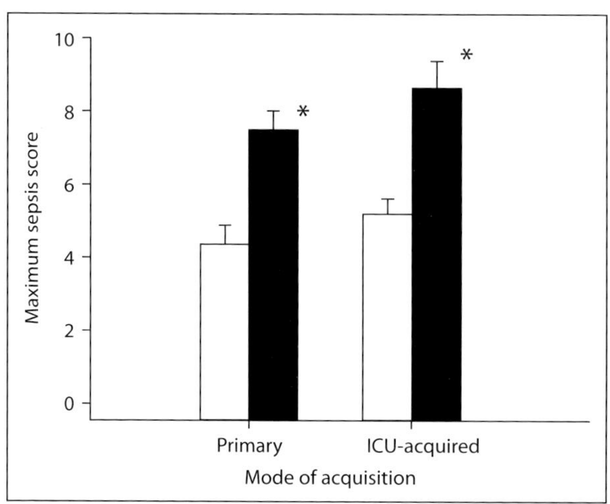

Fig. 4. For patients admitted with infection, and those who developed infection while in the ICU, maximal sepsis scores were significantly higher in nonsurvivors (dark bars) than in survivors (open bars); * $p < 0.01$ [13].

Table 1. Determinants of adverse outcome in patients with clinical sepsis (sepsis score ≥7) [13]

Type of Infection	Survivors	Nonsurvivors	p
Primary, %	38.1	46.7	NS
ICU-acquired, %	76.2	60	NS
None, %	14.3	13.2	NS
Pneumonia, %	47.6	53.3	NS
Peritonitis, %	33.3	40	NS
Bacteremia, %	42.9	20	NS
Sepsis score	7.7±0.2	8.9±0.4	<0.01

Consensus terminology has provided a mechanism for clarifying sepsis as a concept; it has not, however, proven particularly useful in characterizing sepsis as a disease [14]. The delineation of a disease implies not only an understanding of pathologic mechanism, but also the definition of a population of patients who might benefit from specific therapies to modify that process.

How Do Microorganisms Evoke a Response in the Host?

Multicellular organisms such as human beings live in intimate proximity with a complex microbial world. Our mucosal surfaces are colonized by an extraordinarily diverse group of bacteria – current estimates are that the normal flora of the healthy human comprises somewhere between 500 and 1,000 unique species [15]. Microbial cells outnumber host cells by a factor of 10 to 1 [16] and microbial genes outnumber human genes by 100 to 1 [17]. Yet the consequences of this interaction are not only benign, but necessary for normal physiologic development, and one of the more remarkable aspects of the immune system is its capacity not to respond to the indigenous flora.

Yet tissue invasion, or infection, poses a threat to the organism, and a complex response has evolved to counter this threat. That response is based on the capacity of the host to recognize molecular patterns that are foreign to the normal cellular environment and that signal danger [18]. The prototypical mechanism through which danger is recognized involves a family of receptors – encoded in the germ-line – known as Toll-like receptors (TLRs). Distinct TLRs bind and are activated by characteristic molecular patterns [19]. TLR2, for example, binds products from the cell wall of Gram-positive organisms, including lipoteichoic acid and peptidoglycan, while TLR5 is activated by the protein flagellin which is found in bacterial flagellae. TLR9 binds the CpG motifs that are characteristic of bacterial DNA, while TLR3, TLR7 and TLR8 recognize viral nucleic acids. TLR4 is activated by endotoxin from the cell wall of Gram-negative bacteria, as well as by a number of endogenous ligands including oxidized phospholipids [20], elastase [21] and HMGB1 [22].

The consequences of the interaction of a ligand with a TLR are complex, but worthy of consideration as they provide valuable insights into how a response is effected, and therefore how it might be modified therapeutically. Endotoxin or lipopolysaccharide – the focus of this volume – is a prototypical TLR agonist, inducing cellular activation following its engagement with TLR4 (fig. 5). Its interactions with elements of the innate host immune system are more reminiscent of those of a hormone than of a toxin [23].

Endotoxin that has been absorbed into the circulation through the gut or the lung, or shed during the course of an invasive Gram-negative infection, is transported in a complex with a dedicated carrier protein, lipopolysaccharide-binding protein. The lipopolysaccharide-binding protein:endotoxin complex is then capable of being transferred to TLR4, although full activation of the receptor requires the membrane receptor CD14 and an accessory protein, MD2. Engagement of the TLR4 complex results in the recruitment of several adapter proteins, including MyD88, TIRAP and IRAK, creating a signaling complex that, in turn, leads to activation of downstream signaling through the mitogen-activated protein (MAP) kinase and phospahtidylinositol-3 (PI3) kinase pathways, and through the activation of the transcription factor, NF-κB.

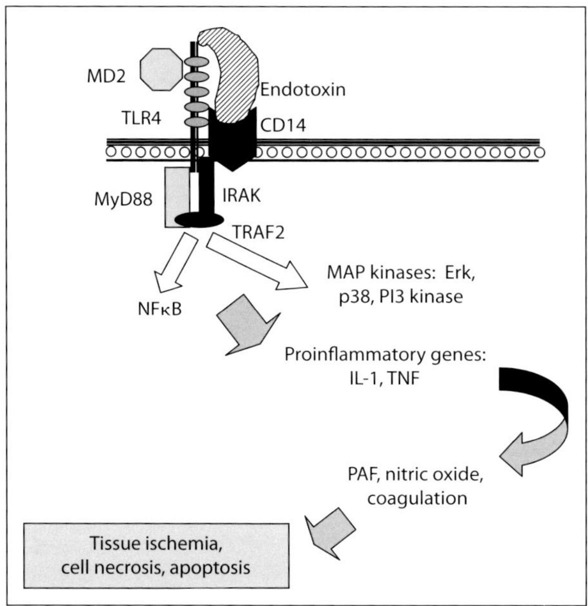

Fig. 5. A schematic representation of the interaction of endotoxin with host cells. Endotoxin activates cells through a dedicated receptor complex, evoking a complex transcriptional response leading to the differential expression of more than 3,700 genes (see text for details).

The consequence is the transcription of members of a family of early proinflammatory genes, including interleukin-1 (IL-1) and tumor necrosis factor (TNF), and their release from the cell. IL-1 and TNF then act on target cells through their own specific receptors, evoking further cellular responses that shape the phenotype of sepsis by, for example, activating coagulation through increased expression of tissue factor or inducing vasodilatation through upregulation of synthesis of nitric oxide catalyzed by inducible nitric oxide synthase. The complexity of the response is underlined by the observation that more than 3,700 genes are either induced or inhibited by exposure to endotoxin in vivo [24].

The biology of Toll-like receptor activation recapitulates observations from clinical studies that the inflammatory response of sepsis is not specific to infection, but a response that can be activated by nonmicrobial endogenous ligands that are abnormally present in an extracellular location. Thus, the endotoxin receptor TLR4 can also be activated by the nuclear protein HMGB1, released from injured cells, or by oxidized phospholipids in the membranes of cells. Indeed the differentiation between infectious triggers is somewhat arbitrary, given the long and intimate interaction between the host and microbial worlds that has occurred over evolutionary history. Mitochondria, for example, owe their evolutionary roots to protobacteria that parasitized primitive unicellular

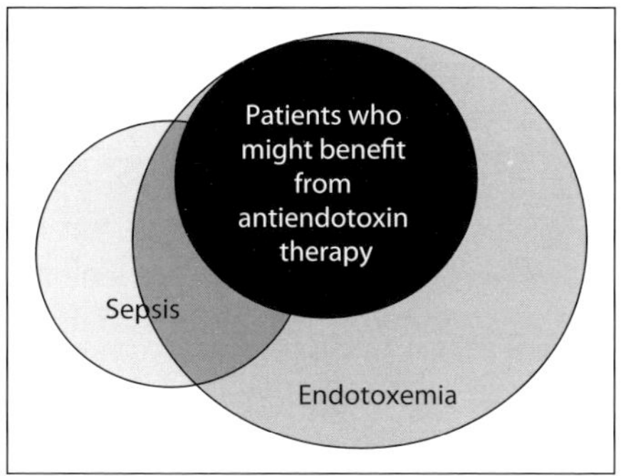

Fig. 6. The relationship between sepsis and endotoxemia. A majority of patients meeting clinical criteria for sepsis can be shown to have elevated levels of circulating endotoxin; however, these elevated levels are also seen in a large number of acutely ill patients who do not meet sepsis criteria. Within this population of endotoxemic patients, it is likely that only a subgroup, as yet unidentifiable, will benefit from reduction of the total endotoxin burden.

organisms more than a billion years ago [25]. This event set the stage for cellular differentiation and the evolution of multicellular organisms. However mitochondria have retained some of the characteristic biochemical features of their ancient origins, and mitochondrial DNA, rich in the CpG motifs found in bacterial DNA, can activate neutrophils by binding to TLR9 [26], providing yet another mechanism through which injured tissue can evoke a response indistinguishable from that resulting from bacterial infection.

At both the clinical and the biochemical level, the distinction between infectious and noninfectious causes of a systemic inflammatory response is arbitrary and inadequate. Attempts to modulate this response therapeutically have proven unsatisfactory, in no small part because sepsis as a concept does not readily define an appropriate patient population for therapeutic intervention. Nowhere is this more apparent than in approaches to treat clinical sepsis by targeting the activity of endotoxin.

Endotoxin and Sepsis: What Is the Connection?

Circulating endotoxin appears to be present in most patients who meet classical clinical criteria for sepsis [27, 28], although several authors have suggested otherwise [29, 30]. This discrepancy reflects, in part, the well-recognized limitations of the Limulus amebocyte lysate assay for endotoxin in protein-containing specimens, but also an important clinical reality: endotoxin is present in some, but not all patients with sepsis, as well as in many patients with acute life-threatening illnesses that would not meet the criteria for sepsis (fig. 6). Moreover, it does not necessarily follow that the simple presence of circulating endotoxin constitutes an adequate rationale for its elimination.

Endotoxemia has been demonstrated in a variety of clinical settings, including following cardiopulmonary bypass [31], in patients with congestive heart failure [32], in chronic renal failure [33], in cirrhosis [34] and in patients with a ruptured abdominal aortic aneurysm [35]. While endotoxemia is prominent in critically ill patients with sepsis, it is also detectable in more than half of all ICU patients on the day of ICU admission, even though the majority of these patients do not meet sepsis criteria [28]. In fact, endotoxemia occurs during strenuous exercise [36], in smokers [37] and following ingestion of a high-fat diet [38]. It is sufficiently ubiquitous during states of physiologic stress that it might readily be considered part of an adaptive response, and not simply an undesirable external insult.

If endotoxemia is a purely pathologic state, then antiendotoxin therapies would be expected to show reproducible evidence of benefit when they are employed in disease processes such as sepsis in which endotoxemia is common. Early studies supported this hypothesis. Ziegler et al. [39], for example, showed that neutralizing endotoxin with an antiserum resulted in improved survival for patients with Gram-negative infections, and particularly for those in shock. A monoclonal antibody directed against endotoxin from a mutant strain of *Escherichia coli* showed similar promise of efficacy in a multicenter study of 543 patients [40], an effect, however, that was not replicated in a subsequent larger study [41]. Similarly the extracorporeal removal of endotoxin using a polymyxin B column has shown evidence of efficacy in pooled data from a number of small trials that recruited an heterogeneous population of patients [42], as well as in a study of patients with severe intra-abdominal infections [43].

There is, however, evidence that neutralization of endotoxemia may not always be beneficial. The lack of efficacy apparent in a number of recent studies of a variety of approaches to neutralize endotoxin in patients with sepsis [44, 45] may be explained by shortcomings of the intervention, suboptimal dosing or a low prevalence of endotoxemia in the target population. It may also reflect the possibility that endotoxemia is an adaptive and beneficial response in some patients with life-threatening infection. Animal studies, for example, show that a genetic inability to respond to endotoxin is associated with an impaired response to infection with *Candida* [46], and patients with a polymorphism in the TLR4 gene show enhanced susceptibility to *Candida* [47]. Intriguingly, neutralization of endotoxin using a monoclonal antibody was associated with an increased risk of death for patients with Gram-positive infection [41].

Elevated levels of circulating endotoxin can cause a syndrome that bears most of the features of clinical sepsis [48], and the acute administration of a large amount can result in organ dysfunction [49]. But endotoxemia, rather than sepsis, is the specific therapeutic target, and the unanswered challenge remains to determine in which patients with endotoxemia will intervention be beneficial.

Conclusions

Elevated circulating levels of bacterial endotoxin are a prominent feature of clinical sepsis, and plausibly linked to the pathogenesis of the resulting morbidity.

From a therapeutic perspective, however, the neutralization of endotoxin can only be beneficial in patients in whom levels are excessive. Our challenge for the future is to redirect our thinking to evaluate antiendotoxin therapies in patients with endotoxemia, rather than in patients with the ill-defined syndrome of sepsis, and then to determine in which of this cohort might endotoxin neutralization be beneficial, as opposed to potentially harmful.

References

1 Bone RC, Balk RA, Cerra FB, et al: Definitions for sepsis and organ failure and guidelines for the use of innovative therapies in sepsis. The ACCP/SCCM Consensus Conference Committee. American College of Chest Physicians/Society of Critical Care Medicine. Chest 1992;101:1644–1655.

2 Angus DC, Linde-Zwirble WT, Lidicker J, Clermont G, Carcillo J, Pinsky MR: Epidemiology of severe sepsis in the United States: analysis of incidence, outcome, and associated costs of care. Crit Care Med 2001; 29:1303–1310.

3 Majno G: The ancient riddle of sigma psi iota sigma (sepsis). J Inf Dis 1991;163:937–945.

4 Capps JA, Coleman GH: Influence of alcohol on prognosis of pneumonia in Cook County Hospital. JAMA 1923;80:750–752.

5 Rogers DE: The changing pattern of life-threatening microbial disease. N Engl J Med 1959;261:677–683.

6 Achievements in public health, 1900–1999. MMWR 1999;48:621–629.

7 Atkins E, Wood WB Jr: Studies on the pathogenesis of fever. II. Identification of an endogenous pyrogen in the blood stream following the injection of typhoid vaccine. J Exp Med 1955;102:499–516.

8 Michalek SM, Moore RN, McGhee JR, Rosenstreich DL, Mergenhagen SE: The primary role of lymphoreticular cells in the mediation of host responses to bacterial endotoxin. J Infect Dis 1980;141:55–63.

9 Wiles JB, Cerra FB, Siegel JH, Border JR: The systemic septic response: does the organism matter? Crit Care Med 1980;8:55–60.

10 Deutschman CS, Konstantinides FN, Tsai M, Simmons RL, Cerra FB: Physiology and metabolism in isolated viral septicemia. Further evidence of an organism independent host dependent response. Arch Surg 1987;122:21–25.

11 Watters JM, Bessey PQ, Dinarello CA, Wolff SM, Wilmore DW: Both inflammatory and endocrine mediators stimulate host responses to sepsis. Arch Surg 1986;121:179–190.

12 Michie HR, Spriggs DR, Manogue KR, et al: Tumor necrosis factor and endotoxin induce similar metabolic responses in human beings. Surgery 1988;104:280–286.

13 Marshall JC, Sweeney D: Microbial infection and the septic response in critical surgical illness. Sepsis, not infection, determines outcome. Arch Surg 1990;125:17–23.

14 Marshall JC: Rethinking sepsis: from concepts to syndromes to diseases. Sepsis 1999; 3:5–10.

15 Guarner F, Malagelada JR: Gut flora in health and disease. Lancet 2003;361:512–519.

16 Savage DC: Microbial ecology of the gastrointestinal tract. Annu Rev Med 1977;31:107–133.

17 Gill SR, Pop M, Deboy RT, et al: Metagenomic analysis of the human distal gut microbiome. Science 2006;312:1355–1359.

18 Matzinger P: The danger model: a renewed sense of self. Science 2002;296:301–305.

19 Miyake K: Innate immune sensing of pathogens and danger signals by cell surface Toll-like receptors. Semin Immunol 2007;19:3–10.

20 Imai Y, Kuba K, Neely GG, et al: Identification of oxidative stress and Toll-like receptor 4 signaling as a key pathway of acute lung injury. Cell 2008;133:235–249.

21 Devaney JM, Greene CM, Taggart CC, Carroll TP, O'neill SJ, McElvaney NG: Neutrophil elastase up-regulates interleukin-8 via toll-like receptor 4. FEBS Lett 2003; 544:129–132.
22 Tsung A, Sahai R, Tanaka H, Nakao A, Fink MP, Lotze MT, et al: The nuclear factor HMGB1 mediates hepatic injury after murine liver ischemia-reperfusion. J Exp Med 2005;201:1135–1143.
23 Marshall JC: Lipopolysaccharide: an endotoxin or an exogenous hormone? Clin Infect Dis 2005;41(Suppl 7):S470–S480.
24 Calvano SE, Xiao W, Richards DR, et al: A network-based analysis of systemic inflammation in humans. Nature 2005;437:1032–1037.
25 Dyall SD, Brown MT, Johnson PJ: Ancient invasions: from endosymbionts to organelles. Science 2004;304:253–257.
26 Zhang Q, Raoof M, Chen Y, et al: Circulating mitochondrial DAMPs cause inflammatory responses to injury. Nature 2010;464:104–107.
27 Opal SM, Scannon PJ, Vincent JL, et al: Relationship between plasma levels of lipopolysaccharide (LPS) and LPS-binding protein in patients with severe sepsis and septic shock. J Infect Dis 1999;180:1584–1589.
28 Marshall JC, Foster D, Vincent JL, et al: Diagnostic and prognostic implications of endotoxemia in critical illness: results of the MEDIC study. J Infect Dis 2004;190:527–534.
29 Danner RL, Elin RJ, Hosseini JM, Wesley RA, Reilly JM, Parrillo JE: Endotoxemia in human septic shock. Chest 1991;99:169–175.
30 Bates DW, Parsonnet J, Ketchum PA, et al: Limulus amebocyte lysate assay for detection of endotoxin in patients with sepsis syndrome. AMCC Sepsis Project Working Group. Clin Infect Dis 1998;27:582–591.
31 Riddington DW, Venkatesh B, Boivin CM, et al: Intestinal permeability, gastric intramucosal pH, and systemic endotoxemia in patients undergoing cardiopulmonary bypass. JAMA 1996;275:1007–1012.
32 Niebauer J, Volk HD, Kemp M, et al: Endotoxin and immune activation in chronic heart failure: a prospective cohort study. Lancet 1999;353:1838–1842.
33 Goncalves S, Pecoits-Filho R, Perreto S, et al: Associations between renal function, volume status and endotoxaemia in chronic kidney disease patients. Nephrol Dial Transplant 2006;21:2788–2794.
34 Lin CY, Tsai IF, Ho YP, et al: Endotoxemia contributes to the immune paralysis in patients with cirrhosis. J Hepatol 2007;46: 816–826.
35 Roumen RMH, Frieling JTM, van Tits HWHJ, et al: Endotoxemia after major vascular operations. J Vasc Surg 1993;18:853–857.
36 Jeukendrup AE, Vet-Joop K, Sturk A, et al: Relationship between gastro-intestinal complaints and endotoxaemia, cytokine release and the acute-phase reaction during and after a long-distance triathlon in highly trained men. Clin Sci (Lond) 2000;98:47–55.
37 Wiedermann CJ, Kiechl S, Dunzendorfer S, et al: Association of endotoxemia with carotid atherosclerosis and cardiovascular disease: prospective results from the Bruneck study. J Am Coll Cardiol 1999;34: 1975–1981.
38 Erridge C, Attina T, Spickett CM, Webb DJ: A high-fat meal induces low-grade endotoxemia: evidence of a novel mechanism of postprandial inflammation. Am J Clin Nutr 2007;86:1286–1292.
39 Ziegler EJ, McCutchan JA, Fierer J, et al: Treatment of Gram-negative bacteremia and shock with human antiserum to a mutant *Escherichia coli.* N Engl J Med 1982;307: 1225–1230.
40 Ziegler EJ, Fisher CJ, Sprung CL, et al: Treatment of Gram-negative bacteremia and septic shock with HA-1A human monoclonal antibody against endotoxin. N Engl J Med 1991;324:429–436.
41 McCloskey RV, Straube RC, Sanders C, Smith SM, Smith CR, the CHESS Trial Study Group: Treatment of septic shock with human monoclonal antibody HA-1A. A randomized double-blind, placebo-controlled trial. Ann Intern Med 1994;121:1–5.
42 Cruz DN, Perazella MA, Bellomo R, et al: Effectiveness of polymyxin B-immobilized fiber column in sepsis: a systematic review. Crit Care 2007;11:R47.

43 Cruz DN, Antonelli M, Fumagalli R, et al: Early use of polymyxin B hemoperfusion in abdominal septic shock: the EUPHAS randomized controlled trial. JAMA 2009;301: 2445–2452.
44 Angus DC, Birmingham MC, Balk RA, et al: E5 murine monoclonal antiendotoxin antibody in Gram-negative sepsis: a randomized controlled trial. E5 Study Investigators. JAMA 2000;283:1723–1730.
45 Dellinger RP, Tomayko JF, Angus DC, et al: Efficacy and safety of a phospholipid emulsion (GR270773) in Gram-negative severe sepsis: results of a phase II multicenter, randomized, placebo-controlled, dose-finding clinical trial. Crit Care Med 2009;37:2929–2938.
46 Netea MG, Van Der Graaf CA, Vonk AG, Verscheueren I, van der Meer JW, Kullberg BJ: The role of Toll-like receptor (TLR) 2 and TLR4 in the host defense against disseminated candidiasis. J Infect Dis 2002;185: 1483–1489.
47 Van Der Graaf CA, Netea MG, Morré SA, et al: Toll-like receptor 4 Asp299Gly/Thr399Ile polymorphisms are a risk factor for Candida bloodstream infection. Eur Cytokine Netw 2006;17:29–34.
48 Lowry SF: Human endotoxemia: a model for mechanistic insight and therapeutic targeting. Shock 2007;24(Suppl 1):94–100.
49 Taveira Da Silva AM, Kaulach HC, Chuidian FS, Lambert DR, Stuffredini AF, Danner RL: Brief report: shock and multiple organ dysfunction after self administration of salmonella endotoxin. N Engl J Med 1993; 328:1457–1460.

John C. Marshall, MD
St. Michael's Hospital, 4th Floor Bond Wing, Rm. 4–007
30 Bond Street
Toronto, ON M5B 1W8 (Canada)
Tel. +1 416 864 5225, Fax +1 416 864 5141, E-Mail marshallj@smh.ca

Ronco C, Piccinni P, Rosner MH (eds): Endotoxemia and Endotoxin Shock: Disease, Diagnosis and Therapy. Contrib Nephrol. Basel, Karger, 2010, vol 167, pp 14–24

Endotoxins and Other Sepsis Triggers

Steven M. Opal

Infectious Disease Division, Memorial Hospital of Rhode Island, Pawtucket, R.I., USA

Abstract

Endotoxin, or more accurately termed bacterial lipopolysaccharide (LPS), is recognized as the most potent microbial mediator implicated in the pathogenesis of sepsis and septic shock. Yet despite its discovery well over a century ago, the fundamental role of circulating endotoxin in the blood of most patients with septic shock remains enigmatic and a subject of considerable controversy. LPS is the most prominent 'alarm molecule' sensed by the host's early warning system of innate immunity presaging the threat of invasion of the internal milieu by Gram-negative bacterial pathogens. In small doses within a localized tissue space, LPS signaling is advantageous to the host in orchestrating an appropriate antimicrobial defense and bacterial clearance mechanisms. Conversely, the sudden release of large quantities of LPS into the bloodstream is clearly deleterious to the host, initiating the release of a dysregulated and potentially lethal array of inflammatory mediators and procoagulant factors in the systemic circulation. The massive host response to this single bacterial pattern recognition molecule is sufficient to generate diffuse endothelial injury, tissue hypoperfusion, disseminated intravascular coagulation and refractory shock. Numerous attempts to block endotoxin activity in clinical trials with septic patients have met with inconsistent and largely negative results. Yet the groundbreaking discoveries within the past decade into the precise molecular basis for LPS-mediated cellular activation and tissue injury has rekindled optimism that a new generation of therapies that specifically disrupt LPS signaling might succeed. Other microbial mediators found in Gram-positive bacterial and viral and fungal pathogens are now appreciated to activate many of the same host defense networks induced by LPS. This information is providing novel interventions in the continuing effots to improve the care of septic patients.

Sepsis and the multiorgan failure that frequently accompanies severe infection remains a leading cause of mortality in the intensive care unit [1]. It is estimated that about 650,000–750,000 patients develop sepsis annually in the United States with similar incidences in Europe and around the world [2]. Nearly half of septic patients develop severe sepsis and septic shock. The mortality for septic shock remains approximately 30–45%, despite advances in supportive care and numerous efforts to improve patient outcome [1–3].

The microbiology of sepsis has significantly evolved over the past 25 years. The principal microbial pathogens in the 1970s were enteric Gram-negative bacilli and *Pseudomonas aeruginosa.* In the late 1980s, a transition to predominantly Gram-positive bacterial pathogens was observed [3]. The rapid transmission and acquisition of antibiotic resistance genes among Gram-positive bacteria, and their propensity to adhere and persist on vascular catheter surfaces and other implantable medical devices have contributed to the increasing incidence of Gram-positive pathogens as a cause of sepsis. Opportunistic fungal pathogens are also increasing in frequency as a cause of sepsis [3].

Remarkably, Gram-negative bacterial pathogens now appear to be staging a comeback as the predominant causative microorganisms of ICU infections in recent surveys [4].

Endotoxin, Microbial Mediators and the Recognition of Sepsis

The consensus working definitions for such clinical terms as sepsis, septic shock, systemic inflammatory response syndrome and multiple organ dysfunction syndrome have been recently updated by the surviving sepsis campaign [2]. These definitions take into account the myriad of infectious agents and microbial mediators implicated in the pathogenesis of sepsis. Actual bloodstream infection by these pathogens at the time sepsis is recognized by the clinician is documented in only about one third of patients, but the evidence of generalized inflammation and procoagulant activity is almost invariably present. The systemic inflammatory response in human sepsis is primarily initiated by microbial-derived, highly conserved, macromolecules that feature surface patterns not found in human tissues. The most potent of all the pathogen-associated molecular pattern (PAMP) molecules is bacterial lipopolysaccharide (LPS), also known as endotoxin. A large number of other PAMPs are expressed on Gram-positive bacteria, fungi, parasites and viral pathogens. These molecules serve as ligands for the pattern recognition receptors expressed on immune effector cells known as the Toll-like receptors (TLRs) [5, 6]. A summary of the major pathogen-derived mediators of sepsis and their respective Toll-like receptors (TLRs) is found in table 1.

The TLR family is the most important, but not the only PAMP recognition receptor complex, within the human innate immune system. TLRs are type 1 transmembrane receptors for the detection of LPS and many other microbial mediators, such as peptidoglycan, lipopeptides, flagellins, microbial nucleic acids, multiple fungal cell wall components, viral proteins and lipoteichoic acid. Ten TLRs have been identified by human genome searches thus far [5].

Table 1. PAMPs and DAMPs (danger-associated molecular patterns) and their primary pattern recognition receptors in humans

	Origin	TLR
Bacterial PAMPs		
LPS-MD2	Gram-negative bacteria	TLR4
Lipoteichoic acid	Gram-positive bacteria	TLR2[a]
Peptidoglycan	Gram-pos./neg. bacteria	TLR2
Triacyl lipopeptides	Gram-pos./neg. bacteria	TLR1/TLR2
Diacyl lipopeptides	*Mycoplasma* spp.	TLR2/TLR6
Porins, OMPs	*Neisseria* spp.	TLR2
Flagellin	motile Gram-pos./neg. bacteria	TLR5
CpG DNA	bacteria, some DNA viruses	TLR9
Viral PAMPs		
dsRNA	double-stranded RNA virus	TLR3
ssRNA	single-stranded RNA virus	TLR7/8
Fungal PAMPs		
Zymosan	*Saccharomyces cerevisiae*	TLR2/TLR6
Phospholipomannan	*Candida albicans*	TLR2
Mannan	*Candida albicans*	TLR4
O-linked mannosyl residues	*Candida albicans*	TLR4
β-glucans	*Candida albicans*	TLR2/dectin-1
DAMPs		
S100a proteins	host	RAGE
Heat shock proteins	host	TLR4
Fibrinogen, fibronectin	host	TLR4
Hyaluronan	host	TLR4
Biglycans	host	TLR4
HMGB1	host	TLR4, TLR2

OMP = Outer membrane protein; CpG = cytosine-phosphate-guanine motifs; RAGE = receptor for advanced glycation endproducts; HMGB1 = high mobility group box-1.
[a] For detection of LTA from some pathogens TLR6 functions as a coreceptor for TLR2.

Microbial Virulence and the Causative Microorganisms of Sepsis

It is important to recognize that most microorganisms lack the requisite capacity to successfully invade humans. Most encounters between microbes and the human immune system results in rapid inhibition and microbial clearance by our innate and adaptive immune systems. Only a select few microbial pathogens possess a highly organized and sophisticated set of virulence properties needed to evade host defenses, invade tissues and detect stress signals within the host. Pathogens also process a series of delivery systems capable of distributing toxins to their cellular targets [7, 8]. These microorganisms have mechanisms for packaging and exchanging favorable gene arrays (e.g. antibiotic resistance genes, virulence genes, pathogenicity islands, repair genes and mutational control genes). This network of virulence genes, known as the 'virulome', work in concert to cause disease if left unchecked by the host's antimicrobial defense mechanisms [8].

The Role of Bacterial Endotoxin

Bacterial LPS is an intrinsic component of the outer membrane of Gram-negative bacteria and is essential for the viability of enteric bacteria. LPS makes up about 75% of the entire outer membrane of enteric bacteria and up to 4 million LPS molecules are found in each bacterial cell wall [8]. The unique potency of endotoxin is illustrated by the recent isolation of an endotoxin-deficient strain of *Neisseria meningitidis,* which is at least 100-fold less potent as an inducer of cytokine production than wild-type bacteria [9]. LPS functions as an alarm molecule, alerting the host at the earliest stage to the possibility of an invasive Gram-negative bacterial infection [10]. LPS release in the circulation provokes a vigorous systemic inflammatory response. It is the host response to LPS, rather than the intrinsic properties of endotoxin itself, that accounts for the potentially lethal consequences attributable to LPS. By comparative analysis, humans, chimpanzees and horses are particularly susceptible to the immunostimulant capacity of LPS, whereas mice and rats are relatively resistant to LPS-mediated toxicity.

LPS is a biphosphorylated, polar macromolecule that usually contains three distinct components: (1) a highly conserved, hydrophobic sequence of fatty acids within its lipid A structure; (2) a less highly conserved, core glycolipid, segment containing some unusual heptose and hexose moieties; and (3) hydrophilic elements expressed on its repeating polysaccharide along its outer surface components [11]. LPS spontaneously forms microaggregates (mini-micelles) in aqueous solutions with its hydrophobic lipid section in the center of the micelle and the hydrophilic polysaccharide components displayed on the outside surface of micelles. In biologic fluids, such as human plasma, LPS rapidly interacts with a variety of serum or membrane-bound lipophilic proteins. Very little LPS circulates freely in the plasma as virtually all LPS molecules are rapidly complexed with circulating proteins and lipoproteins. Three receptors for LPS have been recognized in human cells: (1) soluble or membrane-bound CD14-MD2-TLR4 molecules, (2) CD11/CD18 molecules β_2 integrins) and (3) scavenger receptors for lipid molecules. Soluble and membrane-bound CD14 greatly potentiate the host response to small quantities of LPS and other microbial mediators [11].

In human plasma and other body fluids, LPS trafficking is greatly facilitated by a hepatically derived, acute-phase plasma protein known as LPS-binding protein (LBP). LBP performs as a shuttle molecule picking up polymeric LPS aggregates and transferring LPS monomers to CD14. LPS competes with another neutrophil-derived LPS-binding molecule known as bactericidal/permeability-increasing protein (BPI). Despite BPI's 45% primary amino acid sequence homology with LBP, BPI specifically antagonizes the actions with respect to LPS handling. LBP assists in the delivery of LPS to immune effector cells while BPI inhibits LPS delivery to CD14. The relative concentrations of these two LPS-binding proteins primarily determine the net effect of LPS release [12].

CD14 is a glycosyl phosphatidylinositol-linked protein found primarily on the cell surfaces of myeloid cells. It lacks a transmembrane domain and an intracellular domain and, therefore, is incapable of transducing the LPS signal across cell membranes to activate target cells. After docking to membrane-bound CD14, LPS is then delivered to an essential extracellular adaptor protein known as MD2 (myeloid differentiation factor-2) [10, 13]. The molecular details of LPS binding into the hydrophobic pocket of MD2 are now known in precise ultrastructural detail following the successful crystallization and atomic locations of LPS-MD2 and TLR4 by the work of Park et al. [14]. Hexa-acylated lipid A with precisely aligned 12 or 14 carbon-linked fatty acids fit tightly into the MD2 protein. LPS structures that usually have long carbon-linked fatty acids (C16-C18) or short fatty acids (C8–10) do not fit well into the MD2 pocket and are poor activators of the MD2-TLR4 complex. Likewise, tetra-acylated lipid structures (e.g. eritoran or lipid 4a) occupy the center of MD2, but do not possess the correct surface features to activate TLR4. They act as antagonists to LPS signaling rather than agonists.

The R2-β hydroxyl myristic acid of lipid A is surface exposed and its hydrophobic end aligns into a lipophilic groove along the inner surface of the ectodomain of TLR4 at the interface between the C-terminal domain of TLR4 and the leucine rich repeat loops 15–17 [14]. Once the LPS-MD2 complex is presented to the extracellular domain of TLR4, a large dimeric structure of two LPS-MD2-TLR4 molecule complexes joins together to bring the transmembrane and intracellular domain of the two TLR molecules in close proximity to each other (often as aggregates on lipid rafts along the cell surface). This series of events then engages the necessary adapter molecules with the TIR (Toll interleukin-1 receptor) domain of TLR4, triggering intracellular signaling. The end result of these signal transduction pathways is to activate LPS-responsive gene programs within the nucleus of target cells.

Once TLR4 binds to its LPS ligand, two possible pathways of cellular activation can occur through either the MyD88 (myeloid differentiation factor 88) or the TRIF (Toll-like receptor domain adaptor inducing interferon-β) pathway [6, 10]. A series of signaling events occur with sequential activation of specific tyrosine and threonine/serine kinases. This signaling cascade ultimately leads to phosphorylation, ubiquitylation and degradation of inhibitory κB (IκB) along with other transcriptional activators. IκB degradation releases nuclear factor κB (NFκB) nuclear membrane-binding sites to bind to and translocate into the nucleus. Clotting elements, complement, other acute phase proteins, cytokines, chemokines and nitric oxide synthase genes have NFκB-binding sites at their promoter regions. The outpouring of inflammatory cytokines and other inflammatory mediators after LPS exposure contributes to generalized inflammation, procoagulant activity, tissue injury and septic shock [15, 16].

The Inflammation-Coagulation Networks

Activation of coagulation and generation of a consumptive coagulopathy and diffuse microthrombi are well-recognized complications of severe sepsis (fig. 1). Studies of endotoxin challenge and TNF challenge in normal human volunteers indicate that the extrinsic pathway (tissue factor pathway) is the predominant mechanism by which the coagulation system is activated in human sepsis [17–19]. The contact factors in the intrinsic pathway are also activated and amplify clotting and vasodilation through the generation of bradykinin. Activation of intravascular coagulation results in microthrombi and may contribute to the multiorgan failure that occurs in septic patients. Depletion of coagulation factors and activation of plasmin, antithrombin III and protein C may subsequently lead to a hemorrhagic diathesis as the final manifestation of disseminated intravascular coagulation [18].

The highly interlinked relationship between the coagulation and the innate immune response within the microcirculation has led to numerous attempts to improve the prognosis of sepsis by controlling coagulation [2, 8, 20]. Thrombin and other clotting factors can directly stimulate cytokine and chemokine synthesis in the microcirculation by activating endothelial cells, neutrophils and monocytes via the protease-activated receptors (PAR) [21]. PAR-1 is activated by thrombin and factor X. PAR linkage to its serine protease ligands on endothelial surfaces will increase P-selectin and adhesion molecule expression promoting leukocyte-endothelial cell attachment. This interaction between white cells and the endothelium is advantageous in localized infection for directing phagocytic cells to the site of injury. However, this same system is disadvantageous in generalized inflammation and coagulation in sepsis as diffuse white cells

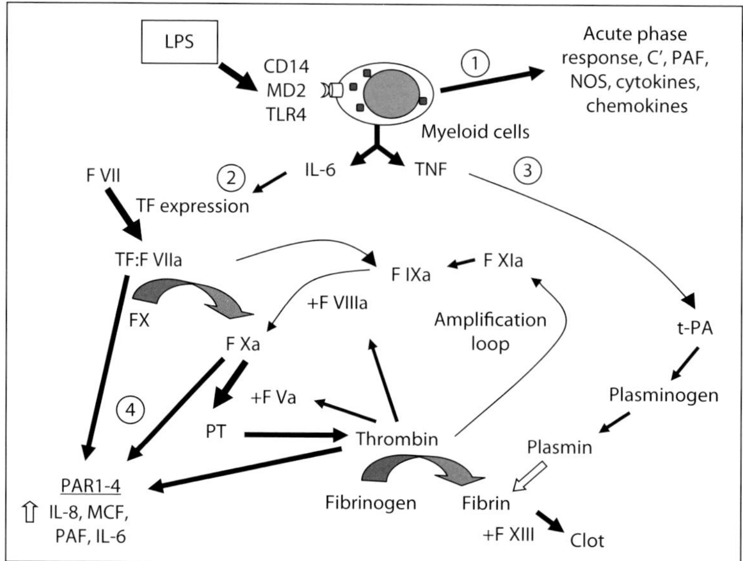

Fig. 1. The endotoxin-coagulation network and interacting signal circuits. (1) LPS recognition by innate immune effector cells initiates acute phase response; (2) interleukin (IL-6)-mediated tissue factor (TF) expression; (3) tumor necrosis factor (TNF)-α-mediated activation of the fibrinolytic system; (4) tissue factor: factor VIIa, FXa and thrombin activation of the PARs (protease activated receptors) 1–4 [TLR; F-factor; t-PA (tissue type plasminogen activator)]. Modified with permission from Opal [26].

binding to endothelial surfaces damage vascular tissues leading to microcirculatory failure [17].

Endotoxin Tolerance (Reprogramming) and Sepsis-Induced Immune Suppression

The phenomenon of endotoxin tolerance (or reprogramming) has been well characterized in experimental models of sepsis and probably also occurs in human sepsis. Endotoxin tolerance is the desensitization to LPS-induced lethality after a priming (small) dose of LPS before an otherwise lethal large dose of LPS. This reprogramming event primarily occurs at the transcriptional level, with downregulation of genes encoding for proinflammatory cytokines and other acute phase proteins. The initial desensitizing dose of endotoxin induces endogenous corticosteroids and anti-inflammatory cytokines such as interleukin-10, decreases cell surface expression of TLRs and major histocompatibility antigens, alters nuclear translocation of signal transduction molecules, and decreases the stability of messenger RNA (mRNA) for cytokine genes [20].

Experimental and clinical evidence indicates that endotoxin reprogramming can be accompanied by a more generalized immunosuppressive state. In addition to these alterations in transcriptional programs, specific subsets of lymphocytes, dendritic cells and epithelial cells undergo apoptosis at an alarmingly high rate in septic patients [8, 20]. Some degree of immune refractoriness is rather commonplace in human sepsis. The widely held view that sepsis is a monopolar hyperinflammatory state is overly simplistic and based upon acute endotoxin models in the animal research laboratory. It is now evident that many patients experience a systemic hypoinflammatory state, often in later phases of septic shock. Repeated insults by microbial mediators result in less profound physiologic alterations than those observed in the initial phases of severe sepsis [8, 20, 22].

Recognition of Other Microbial Mediators by Pattern Recognition Receptors

TLR4 is the primary LPS receptor, whereas TLR2 and its heterodimeric signaling partners, either TLR1 or TLR6, recognize an array of other microbial mediators that serve as PAMPs for fungal, viral, parasitic and Gram-positive bacterial pathogens [10, 13, 22, 23]. Similar to its anchoring role for LPS, CD14 initially binds to bacterial peptidoglycan, lipoteichoic acid and lipopeptides from Gram-positive bacteria and delivers these microbial ligands to TLR2 for intracellular signaling. TLR5 recognizes bacterial flagellin from either Gram-negative or Gram-positive bacteria that possess motility by the action of flagella [5, 15]. The transcriptional profiles and host response patterns to Gram-positive bacteria (lacking LPS) and Gram-negative bacteria significantly differ, indicating that the signaling networks from Gram-positive bacteria are dissimilar from bacteria that possess endotoxin in their outer membrane [24–27].

Specific genetic elements found in bacterial and some viral pathogens are recognized by specialized TLR nucleic acid receptors found within the endosomal compartment inside cells. TLR9 recognizes unmethylated CpG motifs found in bacterial DNA [28], while single-stranded RNA and double-stranded RNA found in viral pathogens are recognized by TLR7/8 and TLR3, respectively [6, 10, 13]. The natural ligand for TLR10 has yet to be identified, but it appears to associate with TLR2 and may form a heterodimeric structure to recognize some microbial ligands [15].

Even if pathogens escape detection and successfully invade the intracellular space, their presence can still be detected by another pattern recognition receptor family known as the NOD-LRR (nucleotide oligomerization domain-leucine rich repeat) proteins. These proteins recognize specific components of Gram-negative and Gram-positive bacterial peptidoglycan, activating acute response genetic programs to eliminate the invader [15, 22]. A separate intracellular recognition system, the RLHs (retinoic acid-inducible gene-1-like helicases) detect intracellular viral genomes and alert the innate immune system to the presence of viral pathogens [22].

Other pattern-recognition molecules include alternative complement components, mannose-binding lectin and CD14 [22, 29]. The innate immune system is by nature an early response and nonspecific antimicrobial defense system. Innate immune function lacks the precision of the adaptive immune system (e.g. B cells and T cells), but the rapid response capability with phagocytosis and clearance of pathogens more than compensates for this lack of precision. The innate immune system, a critical survival mechanism, and its cellular components (neutrophils, monocytes, macrophages and natural killer cells) play a central role in the pathogenesis of septic shock [20].

Other Microbial Toxins in Sepsis

Another important microbial mediator in the pathogenesis of septic shock is bacterial superantigen. Superantigens comprise a diverse group of protein-based exotoxins, from streptococci, staphylococci and other pathogens that all share the capacity to bind to specific sites on major histocompatibility class II molecules on antigen-presenting cells, and activate large numbers of CD4+ T cells, bypassing the usual mechanism of antigen processing and presentation [30].

Whereas a conventional peptide antigen stimulates only about one in 10^5 circulating lymphocytes that can recognize each unique structural epitope, a superantigen (e.g. toxic shock syndrome toxin-1 from *Staphylococcus aureus,* which binds to the Vβ2 region of T cells) can stimulate up to 10–20% of the entire circulating lymphocyte population. This results in excessive activation of both lymphocytes and macrophages, which in turn, leads to the uncontrolled synthesis and release of inflammatory cytokines. Polymicrobial infections with pathogens that release both bacterial superantigens and endotoxin may be particularly injurious to the host. LPS sensitivity is upregulated by superantigens that prime the immune system to subsequent LPS exposure.

Conclusions

The innate immune system is primed to recognize a set of highly conserved molecular patterns that alter the host to foreign invaders. This recognition system is protective against the myriad of minor injuries and infections we all experience over a lifetime. This same alarm system can be potentially lethal if activated in an unregulated and generalized systemic reaction that characterizes the pathophysiology of sepsis. Extracorporeal removal of microbial mediators such as endotoxin, superantigens or other pathogen-derived mediators is a logical therapeutic intervention in septic patients. The tools to successfully remove these injurious mediators are now available. Clinical trials will be needed to determine when to employ these blood purification technologies in caring for septic patients.

References

1 Angus DC, Liinde-Zwirble WT, Lidicker J, Clermont G, Carcillo J, Pinsky MR: Epidemiology of severe sepsis in the United States: analysis of incidence, outcome, and associated costs of care. Crit Care Med 2001;29:1303–1310.

2 Dellinger RP, Levy MM, Carlet JM, Bion J, Parker MM, Jaeschke R, Reinhart K, Angus DC, Brun-Buisson C, Beale R, Calandra T, Dhainaut JF, Gerlach H, Harvey M, Marini JJ, Marshall J, Ranieri M, Ramsay G, Sevransky J, Thompson T, Townsend S, Vender JS, Zimmerman JL, Vincent JL, International Surviving Sepsis Campaign Guidelines Committee, American Association of Critical-Care Nurses, American College of Chest Physicians, American College of Emergency Physicians, Canadian Critical Care Society, European Society of Clinical Microbiology and Infectious Diseases, European Society of Intensive Care Medicine, European Respiratory Society, International Sepsis Forum, Japanese Association for Acute Medicine, Japanese Society of Intensive Care Medicine, Society of Critical Care Medicine, Society of Hospital Medicine, Surgical Infection Society, World Federation of Societies of Intensive and Critical Care Medicine: Surviving Sepsis Campaign: international guidelines for management of severe sepsis and septic shock: 2008. Crit Care Med 2008;36:296–327.

3 Martin GS, Mannino DM, Eaton S, Moss M: The epidemiology of sepsis in the United States from 1979 through 2000. N Engl J Med 2003;348:1546–1554.

4 Opal SM, Calandra T: Antibiotic usage and resistance. gaining or losing ground on infections in critically ill patients? JAMA 2009; 302:2367–2368.

5 Akira S, Takeda K: 2004. Toll-like receptor signaling. Nat Rev Immunol 2004;4:499.

6 Takeda K: Evolution and integration of innate immune recognition systems: the Toll-like receptors. J Endotoxin Res 2005;11:51.

7 Merrell DS, Falkow S: Frontal and stealth attack strategies in microbial pathogenesis. Nature 2004;403:250–256.

8 van der Poll T, Opal SM: Host-pathogen interactions in sepsis. Lancet Infect Dis 2008;8:32–43.

9 Pridmore AC, Wyllie DH, Abdillahi F, Steeghs L, van der Ley P, Dower SK, Read RC: A lipopolysaccharide-deficient mutant of *Neisseria meningitidis* elicits attenuated cytokine release by human macrophages and signals via toll-like receptor (TLR)2 but not via TLR4/MD2. J Infect Dis 2001;183:89–96.

10 Beutler B: Inferences, questions and possibilities in Toll-like receptor signaling. Nature 2004;430:257–263.

11 Opal SM, Scannon P, Vincent J-L, Carroll S, White M, Palardy JE, Parejo N, Pribble JP, Lemke J: Relationship between plasma levels of lipopolysaccharide (LPS) and LPS binding protein in severe sepsis and septic shock. J Infect Dis 1999;180:1584–1589.

12 Opal SM, Marra MN, McKelligan B, Fisher CJ, Palardy JE, Scott R: Relative concentrations of endogenous endotoxin binding proteins in infected body fluids. Lancet 1994;344:429–431.

13 Akira S, Uematsu S, Takeuchi O: Pathogen recognition and innate immunity. Cell 2006;124:783–801.

14 Park BS, Song DH, Kim HM, Choi BS, Lee H, Lee JO: The structural basis of lipopolysaccharide recognition by the TLR4-MD-2 complex. Nature 2009;458:1191–1196.

15 Cristofaro P, Opal SM: Role of Toll-like receptors in infection and immunity: clinical implications. Drugs 2006;66:15–29.

16 Reitsma PH, Branger J, Van Den Blink B, Weijer S, van der Poll T, Meijers JC: Procoagulant protein levels are differentially increased during human endotoxemia. J Thromb Haemost 2003;1:1019–1023.
17 Levi M, Opal SM: Coagulation abnormalities in critically ill patients. Crit Care 2006;10:222–228.
18 Opal SM, Esmon C: Functional relationships between coagulation and the innate immune response and their respective roles in the pathogenesis of sepsis. Crit Care 2003;7: 23–38.
19 Riewald M, Petrovan RJ, Donner A, Mueller BM, Ruf W: Activation of endothelial cell protease activated receptor 1 by the protein C pathway. Science 2002;296:1880–1882.
20 Hotchkiss RS, Karl IE: The pathophysiology and treatment of sepsis. N Engl J Med 2003;348:138–150.
21 Coughlin SR: Thrombin signaling and protease-activated receptors. Nature 2000;407:258–264.
22 Cinel I, Opal SM: Molecular biology of inflammation and sepsis: a primer. Crit Care Med 2009;37:291–304.
23 Levitz SM: Interactions of Toll-like receptors with fungi. Microbes Infect 2004;6:1351–1355.
24 Jenner RG, Young RA: Insights into host responses against pathogens from transcriptional profiling. Nature Rev 2005;3:281–294.
25 Opal SM, Cohen J: Clinical Gram-positive sepsis: does it fundamentally differ from Gram-negative bacterial sepsis. Crit Care Med 1999;27:1608–1616.
26 Opal SM: The host response to endotoxin, anti-LPS strategies and the management of severe sepsis. Int J Med Microbiol 2007;297:365–377.
27 Yu SL, Chen HW, Yang PC, Peck K, Tsai MH, Chen JJW, Lin FY: Differential gene expression in Gram-negative and Gram-positive sepsis. Am J Respir Crit Care Med 2004;169:1135–1143.
28 Dalpke A, Frank J, Peter M, Heeg K: Activation of Toll-like receptor 9 by DNA from different bacterial species. Infect Immun 2006;74:940–946.
29 Hoffmann JA, Kafatos KC, Janeway CA, Ezekowitz RAB: Phylogenetic perspectives in innate immunity. Science 1999;284:1313–1317.
30 Sriskandan S, Ferguson M, Elliot V, Faulkner L, Cohen J: Human intravenous immunoglobulin for experimental streptococcal toxic shock: bacterial clearance and modulation of inflammation. J Antimicrob Chemother 2006;58:117–124.

Steven M. Opal, MD, Professor of Medicine
Infectious Disease Division, Memorial Hospital of R.I.
111 Brewster Street
Pawtucket, RI 02860 (USA)
Tel. +1 401 729 2545, Fax +1 401 729 2795, E-Mail Steven_Opal@brown.edu

Ronco C, Piccinni P, Rosner MH (eds): Endotoxemia and Endotoxin Shock: Disease, Diagnosis and Therapy. Contrib Nephrol. Basel, Karger, 2010, vol 167, pp 25–34

Rationale of Extracorporeal Removal of Endotoxin in Sepsis: Theory, Timing and Technique

Claudio Ronco[a,b] · Pasquale Piccinni[a,b] · John Kellum[c]

[a]Department of Nephrology, Dialysis and Transplantation, San Bortolo Hospital, and [b]International Renal Research Institute (IRRIV), Vicenza, Italy; [c]Department of Critical Care Medicine, University of Pittsburgh, Pittsburgh, Pa., USA

Abstract

Several signs and symptoms in sepsis are due to the presence of endotoxin in the circulation. Both in animal and human models, there is an evident immunological response to the endotoxin insult. Furthermore, altered cardiovascular function, lung dysfunction and acute kidney injury are common in sepsis and endotoxemia. In these circumstances it would be extremely important to identify patients with sepsis in the early phases and to characterize the humoral alterations involved with it, including the identification and quantification of circulating endotoxin. Once this is obtained, it seems logical to try to remove as much of the circulating endotoxin as possible in order to mitigate the clinical effects of this condition. This can be achieved today with a very specific hemoperfusion process utilizing cartridges with immobilized polymixin B in an extracorporeal circuit. This approach seems to provide a significant removal of endotoxin with a significant reduction of its circulating levels. The clinical consequences of this approach can be summarized in a mitigation of the septic cascade in the early phases, with improvement of outcome. Recent clinical results seem to confirm these expectations showing a reduction of mortality in patients with early signs of abdominal sepsis due to recent surgery. This opens a new avenue for intervention in sepsis.

Endotoxemia in Animal and Human Sepsis Models

Recent investigations have greatly advanced our understanding of the human biology of severe inflammation and infection [1]. It is now appreciated that innate immune responses to pathogens occur largely via immunocyte recognition of molecular motifs and the subsequent activation of numerous kinase pathways. It is also evident that temporally overlapping pro- and anti-inflammatory signals are generated during such responses. Unfortunately, variability of the clinical phenotype frequently precludes real-time interpretation of such influences and limits opportunities for effective intervention.

The initial human response to infectious challenge from the elective administration of endotoxin, a lipopolysaccharide (LPS) which is a major component of the Gram-negative bacteria outer membrane, has been elucidated in several studies. Within 1 h after the intravenous administration of endotoxin (LPS), subjects may variably experience symptoms including chills, headache, back pain, myalgias, nausea and photophobia. The expression of discomfort from these symptoms varies between subjects, although virtually all subjects report attenuation of symptoms within 4–6 h.

The most reproducible features include increases in core temperature (1–4°C) and heart rate. These manifestations of systemic inflammation generally decrease within 6–8 h [2, 3]. The appearance of cytokines follows a characteristic pattern, beginning with tumor necrosis factor (TNF) activity, which is found to be increased in the circulation within 45–60 min after LPS injection. Endotoxemia elicits dynamic and reproducible changes in the circulating leukocyte population as well as the function of such cells. The circulating leukocyte count begins to decline within 15–30 min after LPS challenge [4] at a time before the onset of clinical symptoms or mediator appearance. This is noteworthy for a marked monocytopenia that reverts toward normal over the ensuing 6–8 h.

A progressive decline in total lymphocyte count also begins shortly after LPS challenge and continues over the ensuing 9–12 h. The differential of the leukocyte count is influenced largely by the circulating polymorphonuclear leukocyte population that, after an early decline, rises to levels above basal within 4–6 h and returns to normal within 24 h. Endothelial cells can be included among the immune cell populations exhibiting activation after in vivo LPS challenge. As evidenced by surrogate markers, such as soluble E-selectin, robust endothelial cell responses can be observed within 2 h of LPS challenge and persist for over 6 h afterward.

Cardiovascular Response in Sepsis

Marked abnormalities in cardiovascular function accompany septic shock, and bacterial endotoxin is believed to be one of the principal mediators of these abnormalities [5]. As empirically demonstrated, the administration of endotoxin to normal subjects causes a depression of left ventricular function that is independent of changes in left ventricular volume or vascular resistance. The changes in function are similar to those observed in septic shock and suggest that endotoxin is a major mediator of the cardiovascular dysfunction in this condition.

Septic shock is characterized by myocardial dysfunction, vasoplegia and microvascular thromboses leading to multiple organ dysfunction and death. The marked cardiac depression witnessed in human clinical sepsis has been simulated in numerous experimental systems. The administration of LPS to human volunteers results in a septic-like syndrome accompanied by decreased ventricular ejection fractions, biventricular dilatation and altered cardiac index [6]. LPS may exert its effects by directly acting on cells, but also via downstream mediators including cytokines, adhesion molecules, nitric oxide and reactive oxygen species. Kumar et al. [7] demonstrated that LPS-induced myocardial dysfunction is mediated by TNF and interleukin-1b (IL-1b), although other downstream mediators have been implicated. Others [8] have documented an increase of myocardial TNF after LPS stimulation, partly synthesized locally by cardiac myocytes themselves. Local myocardial TNF levels may be an important factor in the progression of myocardial dysfunction because TNF both suppresses myocardial contractility and induces cardiac myocyte apoptosis [9].

Pulmonary Response in Sepsis

The lung is particularly susceptible to acute injury in shock. This injury may progress to life-threatening adult respiratory distress syndrome [10]. The ability of endotoxin to induce the release of IL-1 and TNF by both circulating monocytes and the liver suggests that endotoxin may also stimulate local pulmonary production of these inflammatory mediators by alveolar macrophages either directly or through secondary mediators. Following their release, IL-1 and TNF could act on alveolar capillaries in a manner similar to their described action on systemic vessel endothelium, thereby promoting increased alveolar permeability and secondary fluid accumulation. In this regard, the persistence of lung dysfunction after the systemic changes of septic shock have subsided suggests that local factors are indeed important in shock lung.

Renal Response in Sepsis and Sepsis-Related Acute Kidney Injury

The kidney is a target organ in sepsis with significant alterations in tissue and function. Acute kidney injury related to sepsis has been closely linked to renal cell apoptosis [11]. In Gram-negative sepsis, LPS can directly cause apoptosis of tubular cells through Fas-mediated and caspase-mediated pathways, and increased plasma levels of soluble Fas has been described in septic patients [12]. Additionally, experimental models of sepsis have shown that increased caspase activation is associated with the presence of acute renal failure [13].

Apoptosis, an energy-dependent process whereby cells carry out programmed death, contributes to the pathogenesis of acute renal failure. Recently,

Jo et al. [11] and Bordoni et al. [14] suggested that Fas-mediated and caspase-mediated apoptosis of tubular cells might be one of the possible mechanisms involved in endotoxemia-induced renal dysfunction. Consistent with these findings, several studies have shown that circulating LPS may cause an inappropriate activation of proapoptotic pathways in immune, epithelial and endothelial cells [15]. Moreover, LPS can directly act on kidney-resident cells such as podocytes and tubular epithelium, stimulating the synthesis of inflammatory mediators. An interesting rationale emerges to attempt removal of circulating LPS with extracorporeal therapies and a consequent inactivation of circulating proapoptotic factors to prevent kidney damage [16]. Of course such therapy should be attempted well before renal function impairment occurs.

Endotoxemia Is an Important Factor of Morbidity and Mortality in Sepsis

Although endotoxin has generally been presumed to be the major bacterial toxin responsible for septic shock syndrome, the detection of circulating endotoxin in human sepsis has correlated inconsistently with both bacteremia and important clinical endpoints [17]. Almost three quarters of clinically diagnosed cases of septic shock with detectable endotoxemia do not have documented Gram-negative infections. This suggests that endotoxin can persist or possibly increase in the circulation after host defenses or antibiotics, or both, have rendered blood cultures negative. Furthermore, extravascular sources of endotoxin, such as the gastrointestinal tract or sequestered foci of infection, may be important in causing detectable endotoxemia in some patients. In contrast, not all patients with Gram-negative bacteremia are found to be endotoxemic. This well-described phenomenon [18] probably is attributable to low levels of bacteremia with rapid clearance of any free endotoxin from the circulation. Thus, the source of endotoxemia may be undetected Gram-negative bacteremia or, alternatively, release of endotoxin from extravascular sites. In contrast to what has been observed for endotoxemia, positive blood cultures or the presence of Gram-negative bacteremia were not associated with any measures of disease severity or predictive of outcome. Septic shock in humans produces characteristic cardiovascular changes including a low systemic vascular resistance, high cardiac output, a dilated ventricle and a depressed left ventricular ejection fraction. The fall in ejection fraction typically occurs in the first two days after the onset of sepsis and is reversible in those who survive more than 7–10 days.

In patients with septic shock, endotoxemia is generally associated with important laboratory, cardiovascular and clinical consequences of sepsis including lactic acidemia, myocardial dysfunction, organ failure and death. Previous studies have shown that persistent hypotension (due to a low systemic vascular resistance or severe myocardial depression, or both) or multiple organ system failure are the most frequent causes of death in septic shock [19]. These associations provide strong evidence that endotoxin is an important mediator of the high morbidity and mortality of septic shock.

Plasma Levels of LPS and LPS-Binding Protein

Plasma levels of LPS and LPS-binding protein (LBP) represent an important finding in sepsis and septic shock. In a study from Opal et al. [20], where all patients met consensus definitions of severe sepsis, 80% were found to be in septic shock. The average APACHE II score was 26 ± 13.6, and the 28-day all-cause mortality rate for the entire study group (n = 253) was 32.4%. Severe sepsis patients with elevated endotoxin levels had significantly greater mortality rates than those patients without measurable endotoxin. In some studies the 28-day all-cause mortality was 35% in endotoxemic patients and only 22% in nonendotoxemic patients. The greater quartiles of plasma endotoxin levels were associated with shorter survival times ($p < 0.05$). Detectable levels of endotoxin in the plasma were more common in patients with shock than in patients without shock. No correlation was observed between the endotoxin levels and LBP levels found in the systemic circulation of these patients.

There was also no significant association between the type of microorganism that caused the septic process and either plasma endotoxin or LBP levels. In general, patients with documented Gram-negative bacterial sepsis had similar levels of endotoxin and LBP as did patients with Gram-positive bacterial sepsis or fungal sepsis. This study shows that endotoxin is frequently found in the systemic circulation in the presence of sepsis, regardless of the infecting microorganism. Endotoxemia may have originated from unrecognized Gram-negative infections in some patients or from enteric bacteria within the gastrointestinal tract [21]. Regional hypoperfusion and mucosal ischemia are thought to promote the translocation of endotoxin from the intestinal lumen to the systemic circulation [21, 22].

Infusion of endotoxin in animals and in healthy human hosts activates a signaling cascade analogous to sepsis, which, at high levels, results in organ dysfunction. However, the degree of organ dysfunction does not show a direct dose-response relationship with the amount of LPS infused [23]. Genetic background and sex have been shown to influence in vivo human responses to endotoxin challenge [24]. In clinical testing, levels of endotoxin on admission to intensive care units (ICUs) have been shown to be predictive of the development of organ dysfunction in patients with sepsis [25]. Endotoxemia is common in critically ill patients and levels of endotoxin have been reported to vary over time [22, 26].

Endotoxin levels in the normal host are tightly regulated through a highly conserved series of mechanisms for endotoxin binding, signaling and clearance. Transient endotoxemia has been found in many healthy host states, including marathon runners, Olympic athletes and smokers [27, 28]. However, in disease, the mechanisms that regulate endotoxin may vary as the inflammatory response becomes uncoupled from the inciting injury. The breakdown of these regulatory mechanisms along with ongoing translocation of gut-derived endotoxin may contribute to the fluctuating levels observed in several patients. Other potential contributors to fluctuating levels of endotoxin can also include an uncontrolled or recurrent source of Gram-negative sepsis.

The concept of linking recurrent exposure to endotoxin to the overwhelming, uncontrolled systemic inflammatory response associated with multiple organ failure in critical illness is attractive. If ongoing fluctuations in endotoxin levels play a role not only as a marker of ongoing injury, but also as a mediator, it may be possible to measure and modulate this process. A broader understanding of the kinetics of changing levels of endotoxin over time in critical illness may improve therapeutic timing and targeting of specific antiendotoxin therapies [29].

Diagnostic and Prognostic Implications of Endotoxemia in Critical Illness

Although endotoxin is ubiquitous, it has been notoriously difficult to measure reliably in human illness. The most commonly used diagnostic test – the chromogenic limulus amebocyte lysate assay – has been widely used to detect endotoxin contamination of drugs and fluids; however, its utility in biological samples has been limited [30] due to circulating inhibitors of the coagulation reaction. Moreover, other microbial products, notably from fungi, can activate the limulus reaction. Therefore, the assay is not specific for endotoxin. Gram-negative infection is one of the many infectious causes. Although microbiologically proven infection is relatively uncommon in cohort studies, endotoxemia is generally present on the day of ICU admission in the majority of patients.

In a cohort study, Marshall et al. [25], by means of the new Endotoxin Activity (EA) Assay (approved by the US Food and Drug Administration) found EA levels ≤0.40 in 367 patients (42.8% of the population), between 0.40 and 0.60 in 228 patients (26.6%), and ≥0.60 in 262 patients (30.6%); thus, the majority of patients of the study had elevated endotoxin levels at the ICU admission. Criteria for severe sepsis were present at the time of admission in 74 patients (8.6% of the study cohort). ICU mortality for this population was higher than that of patients who did not have sepsis (32.4 vs. 11.5%; $p < 0.001$). EA levels were significantly higher for patients who met criteria for severe sepsis (0.57 ± 0.26 vs. 0.46 ± 0.26 units; $p < 0.001$), and the risk of severe sepsis increased with increasing increments of EA levels. Patients with endotoxemia had increased ICU and hospital mortality. The ICU length of stay was modestly longer for patients with EA levels ≥0.60 at the time of ICU admission (6.8 ± 12.2 vs. 4.9 ± 10.7 days; $p < 0.05$). However, patients who had intermediate or high levels of endotoxin on the day of admission were clearly a sicker population, as reflected by higher admission APACHE II scores and a greater prevalence of severe sepsis. Moreover, patients with the highest levels of circulating endotoxin had a significantly increased risk of dying while in the ICU. Thus, the presence of endotoxemia identifies a high-risk subpopulation of critically ill patients.

Whether this increased mortality risk might be reduced by specific measures to neutralize endotoxin is unknown, but the hypothesis is an attractive one. Not only have the results of animal studies supported a pathogenic role for increased concentrations of endotoxin, but previous analyses of studies designed to neutralize endotoxin in human sepsis have suggested that the greatest potential for benefit occurs in those patients in whom endotoxemia is present, whereas intervention may actually harm those with infection caused by non-Gram-negative organisms [31].

A Targeted Extracorporeal Therapy for Endotoxemia

Since its discovery, whatever the source, endotoxemia is associated with increased organ dysfunction and risk of death in critically ill patients [25]. Furthermore, blood levels of endotoxin have been reported to vary over time in patients, suggesting subsequent waves of exposure either from infection or from intestinal translocation. However, despite new methods of detecting endotoxin, treatment is quite limited. Bacterial infection is treated with antibiotics, but there is no effective method of restoring gut barrier function. Furthermore, many antibiotics result in endotoxin release as bacteria are killed.

Antiendotoxin therapies, like antiendotoxin antibody (HA-1A or E5), have been disappointing, and have failed to demonstrate benefit in patients with confirmed Gram-negative sepsis. Polymyxin B is a well-known antibiotic that has high affinity for endotoxin and is able to neutralize it, although it is associated with neurotoxicity and nephrotoxicity, which precludes its systemic use. Since 1994, polymyxin B has been bound and immobilized to polystyrene fibers. When it is used for hemoperfusion treatment, it can effectively bind endotoxin both in vitro and in vivo. This therapy has been available in Japan since 1994 and thousands of patients have been treated. Unfortunately, despite widespread use in Japan, no large randomized trials have established the efficacy of polymyxin B hemoperfusion.

Several small studies have been conducted, however, and most have studied similar patients. In a recent systematic review, Cruz et al. [32] identified a total of 28 publications, including randomized controlled trials, of polymyxin B hemoperfusion for treatment of sepsis and septic shock. Polymyxin B hemoperfusion was associated with a significantly lower mortality compared to conventional therapy (relative risk: 0.53, 95% CI: 0.43–0.65). Secondary endpoints such as mean arterial pressure increase, vasopressor decrease and mean partial pressure of oxygen in arterial blood/forced inspiratory pressure of oxygen ratio increase were also highly significant. Nevertheless, polymyxin B hemoperfusion appears to have favorable effects on survival and hemodynamics, and the authors argue for the need for further rigorous study of this therapy.

More recently, the same authors reported the results of a multicenter randomized controlled trial on a population of patients with septic shock arising from abdominal infection [33]. The authors reported for the first time prospectively a superior outcome in patients treated with two sessions of hemoperfusion with polymyxin B cartridges compared to patients treated with the current standard of care. This study opens new avenues for the evaluation of this therapeutic approach in other selected populations with endotoxin-mediated septic shock. Indeed, given the poor overall outcomes associated with endotoxemia, polymyxin B hemoperfusion would seem to be a welcome intervention, particularly now that better methods for detection of endotoxemia have become available. Thus, the time is right to confirm the encouraging results recently published in the literature with adequately powered trials and clinical registries combining diagnostic and therapeutic interventions on endotoxin–mediated syndromes.

The theory is in favor of the use of extracorporeal therapies where other therapeutic approaches have failed. The rationale is fully consistent with the experimental findings in animal and human models. The technology is simple and safe and ready to be utilized on a larger scale. All the ingredients for a possible success are included in this new recipe.

References

1 Lowry SF: Human endotoxemia: a model for mechanistic insight and therapeutic targeting. Shock 2005;24(Suppl 1):94–100.

2 Suffredini AF, Shelhamer JH, Neumann RD, Brenner M, Baltaro RJ, Parillo JE: Pulmonary and oxygen transport effects of intravenously administered endotoxin in normal humans. Am Rev Respir Dis 1992;145:1398–1403.

3 Smith PD, Suffredini AF, Allen JB, Wahl LM, Parillo JE, Wahl SM: Endotoxin administration to humans primes alveolar macrophages for increased production of inflammatory mediators. J Clin Immunol 1994;14:141–148.

4 Richardson RP, Rhyne CD, Fong Y, Hesse DG, Tracey KJ, Marano MA, Lowry SF, Antonacci AC, Calvano SE: Peripheral blood leukocyte kinetics following in vivo lipopolysaccharide (LPS) administration to normal human subjects: influence of elicited hormones and cytokines. Ann Surg 1989;210:239–245.

5 Suffredini AF, Fromm RE, Parker MM, Brenner M, Kovacs JA, Wesley RA, Parrillo JE: the cardiovascular response of normal humans to the administration of endotoxin. N Engl J Med 1989;321:280–287.

6 Aras O, Shet A, Bach RR, Hysjulien JL, Slungaard A, Hebbel RP, Escol G, Jilma B, Key NS: Induction of microparticle- and cell-associated intravascular tissue factor in human endotoxemia. Blood 2004;103: 4545–4553.

7 Kumar A, Thota V, Dee L, Olson J, Uretz E, Parillo JE: Tumor necrosis factor-alpha and interleukin 1-beta are responsible for in vitro myocardial cell depression induced by human septic shock serum. J Exp Med 1996;183:949–958.

8 Giroir BP, Johnson JH, Brown T, Allen GL, Beutler B: The tissue distribution of tumor necrosis factor biosynthesis during endotoxemia. J Clin Invest 1992;90:693–698.
9 Meldrum DR: Tumor necrosis factor in the heart. Am J Physiol 1998;274:R577–R595.
10 Smith PD, Suffredini AF, Allen JB, Wahl LM, Parillo JE, Wahl SM: Endotoxin administration to humans primes alveolar macrophages for increased production of inflammatory mediators. J Clin Immunol 1994;14:141–148.
11 Jo SK, Cha DR, Cho WY, Kim HK, Chang KH, Yun SY, Won NH: Inflammatory cytokines and lipopolysaccharide induce Fas-mediated apoptosis in renal tubular cells. Nephron 2002;91:406–415.
12 Guo R, Wang Y, Minto AW, Quigg RJ, Cunningham PN: Acute renal failure in endotoxemia is dependent on caspase activation. J Am Soc Nephrol 2004;15:3093–3102.
13 Wesche-Soldato DE, Swan RZ, Chung CS, Ayala A: The apoptotic pathway as a therapeutic target in sepsis. Curr Drug Targets 2007;8:493–500.
14 Bordoni V, Bolgan I, Brendolan A, Crepaldi C, Gastaldon F, D'Intini V, Pilotto L, Inguaggiato P, Bonello M, Galloni E, Everard P, Bellomo R, Ronco C: Caspase-3 and -8 activation and cytokine removal with a novel cellulose triacetate superpermeable membrane in an in vitro sepsis model. Int J Artif Organs 2003;26:897–905.
15 Camussi G, Mariano F, Biancone L, De Martino A, Bussolati B, Montrucchio G, Tobias PS: Lipopolysaccharide binding protein and CD14 modulate the synthesis of platelet-activating factor by human monocytes and mesangial and endothelial cells stimulated with lipopolysaccharide. J Immunol 1995;155:316–324.
16 Cantaluppi V, Assenzio B, Pasero D, Romanazzi GM, Pacitti A, Lanfranco G, Puntorieri V, Martin EL, Mascia L, Monti G, Casella G, Segoloni GP, Camussi G, Ranieri VM: Polymyxin-B hemoperfusion inactivates circulating proapoptotic factors. Intensive Care Med 2008;34:1638–1645.
17 Stumacher RJ, Kovnat MJ, McCabe WR: Limitations of the usefulness of the limulus assay for endotoxin. N Engl J Med 1973;283: 1261–1264.
18 Jorgensen JH: Clinical applications of the limulus amebocyte lysate test; in Proctor RA (ed): Clinical Aspects of Endotoxin Shock: Handbook of Endotoxin. Amsterdam, Elsevier, 1986, vol 4, pp 127–160.
19 Herring WB, Herion JC, Walker RI, Palmer JG: Distribution and clearance of circulating endotoxin. J Clin Invest 1963;43:79–87.
20 Opal SM, Scannon OJ, Vincent JL, White M, Carroll SF, Palardy JE, Parejo NA, Pribble JP, Lemke JH: Relationship between plasma levels of LPS and LPS-binding protein in patients with severe sepsis and septic shock. J Infect Dis 1999;180:1584–1589.
21 Hurley JC: Endotoxemia: methods of detection and clinical correlates. Clin Microbiol Rev 1995;8:268–292.
22 Danner RL, Elin RJ, Hosseini JM, Wesley RA, Reilly JM, Parillo JE: Endotoxemia in human septic shock. Chest 1991;99:169–175.
23 Stephens RC, O'Malley CM, Frumento RJ, et al: Low-dose endotoxin elicits variability in the inflammatory response in healthy volunteers. J Endotoxin Res 2005;11:207–212.
24 Coyle SM, Calvano SE, Lowry SF: Gender influences in vivo human responses to endotoxin. Shock 2006;26:538–543.
25 Marshall JC, Foster D, Vincent JL, et al: Diagnostic and prognostic implications of endotoxemia in critical illness: results of the MEDIC study. J Infect Dis 2004;190:527–534.
26 Klein DJ, Derzko A, Foster D, Seely AJE, Brunet F, Romaschin AD, Marshall JC: Daily variation in endotoxin levels is associated with increased organ failure in critically ill patients. Shock 2007;28:524–529.
27 Brock-Utne JG, Gaffin SL, Wells MT, et al: Endotoxaemia in exhausted runners after a long-distance race. S Afr Med J 1988;73: 533–536.
28 Hasday JD, Bascom R, Costa JJ, et al: Bacterial endotoxin is an active component of cigarette smoke. Chest 1999;115:829–835.
29 Opal SM, Gluck T: Endotoxin as a drug target. Crit Care Med 2003;31(Suppl 1):57–64.
30 Cohen J: The detection and interpretation of endotoxemia. Intensive Care Med 2000; 26(Suppl 1):S51–S56.

31 Ziegler EJ, Fisher CJ, Sprung CL, et al: Treatment of Gram-negative bacteremia and septic shock with HA-1A human monoclonal antibody against endotoxin. A randomized, double-blind, placebo-controlled trial. The HA-1A Sepsis Study Group. N Engl J Med 1991;324:429–436.

32 Cruz DN, Perazella MA, Bellomo R, de Cal M, Polanco N, Corradi V, Lentini P, Nalesso F, Ueno T, Ranieri VM, Ronco C: Effectiveness of polymyxin B-immobilized fiber column in sepsis: a systematic review. Critical Care 2007;11:R47.

33 Cruz DN, Antonelli M, Fumagalli R, et al: Early use of polymyxin B hemoperfusion in abdominal septic shock: the EUPHAS randomized controlled trial. JAMA 2009;301: 2445–2452.

Claudio Ronco, MD
Department of Nephrology, Dialysis and Transplantation, San Bortolo Hospital
Viale Rodolfi 37
IT–36100 Vicenza (Italy)
Tel. +39 0 444993869, Fax +39 0 444993949, E-Mail cronco@goldnet.it

Ronco C, Piccinni P, Rosner MH (eds): Endotoxemia and Endotoxin Shock: Disease, Diagnosis and Therapy. Contrib Nephrol. Basel, Karger, 2010, vol 167, pp 35–44

Extracorporeal Removal of Endotoxin: The Polymyxin B-Immobilized Fiber Cartridge

Tohru Tani[a] · Hisataka Shoji[b] · Gualtiero Guadagni[c] · Angelo Perego[d]

[a]Department of Surgery, Shiga University of Medical Science, Otsu City, [b]Division of Emergency and Critical Care Medicine, Toray Medical Co., Ltd., Tokyo, Japan; [c]ESTOR S.p.A., Milan, [d]U.O.C Nefrologia e Dialisi, Ospedale di Monselice, Monselice, Italy

Abstract

Endotoxin, which consists of lipopolysaccharide (LPS), is an outer membrane component of the Gram-negative bacterial cell wall. Endotoxin in the blood stream from an infectious focus or through translocation from the gut plays an important role in the pathogenesis of severe sepsis and septic shock. It binds to monocytes and macrophages, activating them to trigger the production of a variety of mediators. These mediators injure endothelial cells and induce microcirculatory dysfunction. This leads to subsequent organ dysfunction and multiorgan failure. The neutralization or elimination of endotoxin in the blood is an enticing approach for treating severe sepsis and septic shock. Selective adsorbent therapy targeting blood endotoxin has been clinically applied for more than 15 years, mainly in Japan and more recently in Italy and other countries. Toraymyxin™ (PMX; Toray, Tokyo, Japan) is a selective blood endotoxin removal cartridge. PMX is composed of polymyxin B (PL-B) covalently bonded to polystyrene-derivative fibers. It is well known that PL-B binds endotoxin and has bactericidal activity. PL-B has a strong affinity to endotoxin and is able to bind the lipid A portion of endotoxin through ionic and hydrophobic interactions. Intravenous injection of PL-B has significant nephrotoxic and neurotoxic effects. However, covalently immobilized PL-B on the adsorbents of PMX do not leak out into the blood stream, thus allowing the clinical application without the known toxic effects of PL-B. Within each cartridge, an adsorbent structure made of PL-B-fixed fabrics is included. Blood flow direction is well controlled by adopting a radial flow system. PMX treatment occurs by hemoperfusion at a blood flow rate of about 80–120 ml/min. Heparin is preferably used as an anticoagulant. In Japan, PMX has been clinically used since 1994 under the national health insurance system. It is estimated that over 80,000 patients have received PMX treatment in Japan. Not only has PMX been clinically used safely in Japan, but also in other countries.

Bacterial lipopolysaccharide (LPS or endotoxin) is one of the most powerful known molecules of bacterial signaling. Animals and humans have sensitive mechanisms for recognizing the presence of LPS in their tissues. Monocytes, macrophages and neutrophils, expressing specific receptors, the transductors CD14 and Toll-like receptor-4, are particularly sensitive to the presence of LPS. They respond by producing inflammatory mediators which enhance and diversify the LPS signal, triggering the host defenses which isolate and destroy the invading bacteria.

For reasons not entirely clear, the response to the LPS can be excessive and may lead to severe sepsis, septic shock and even death.

In the mid-1970s, polymyxin B (PL-B) was discovered to be protective against endotoxin-induced hemodynamic shock but, at the same time, was demonstrated to be extremely toxic for the kidney and central nervous system. Based on this, several antiendotoxin strategies have been proposed (e.g. monoclonal antibodies, antiendotoxin vaccines, inhibition of endotoxin synthesis); however, those failed in demonstrating reproducible outcomes in septic subjects. In 1983 Toray Industries Inc. developed a blood endotoxin removal cartridge, Toraymyxin™ (PMX), which could be clinically applied by direct hemoperfusion. PMX consists of a polystyrene-based, fibrous adsorbent in which the antibiotic PL-B is covalently immobilized as a ligand to adsorb endotoxin. PMX was approved for Japan in 1994, CE-marked in 1998 and has been commercially available in Europe since 2002. PMX is the only device currently available on the market for the removal of endotoxins from blood. The working principle is based on the antiendotoxin potential of PL-B. Since PL-B is known to have dramatic side effects for the organism when used systemically, this molecule is linked to an inert material in the PMX cartridge, with a covalent bond to prevent its release into the bloodstream.

Endotoxins and Polymyxin B

Endotoxin invasion in the organism is followed by an immediate reaction of monocytes which bind it through the antigenic portion and start the inflammatory reaction. The monocyte activation is followed by the intervention of lipoproteins which also bind free endotoxin through the lipid A portion and try to steal LPS from monocytes in the global attempt to fight the increase of the inflammatory state [1].

If the endotoxin level is high, the monocyte response exceeds the lipoprotein-mediated LPS clearance, creating the conditions in which sepsis can originate. Therefore, in order to remove endotoxin from whole blood, a strong binding force is needed, as LPS which has already bonded to monocytes or lipoprotein complexes has to be removed.

PL-B is an antibiotic agent that has strong bactericidal activity for Gram-negative bacilli, and also has a strong affinity with endotoxin. PL-B plays a

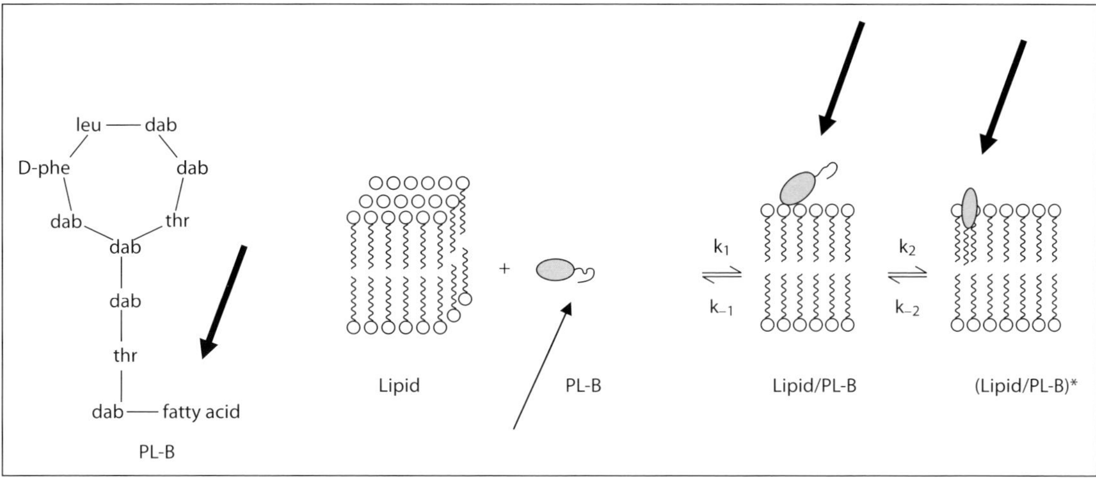

Fig. 1. Interaction between the hydrophobic residues of the PL-B and the acid chains of LPS. The hydrophobic residue is indicated by arrows.

double role towards endotoxin: binding and neutralization. The neutralization is the most important action and depends on the interaction between the hydrophobic residues of PL-B and the fatty acid chains of LPS. The results of this chemical interaction is a unimolecular compound (fig. 1).

How to Design the PL-B-Immobilized Fiber and Cartridge

PL-B cannot be intravenously injected due to its nephrotoxicity and neurotoxicity. Therefore, it was covalently bonded to an insoluble substrate as a ligand and used as a selective adsorbent for an endotoxin. Polystyrene- and polypropylene-conjugated fibers, with island-sea type-conjugated fibers and polypropylene (island component) to provide reinforcement to the fibers, were utilized as a substrate fiber. A bundle of the conjugated fibers were knitted into a fabric and the polystyrene component of the fibers was chemically modified to introduce α-chloroacetamide methyl groups as functional groups to fix PL-B. PL-B has 5 primary amino groups derived from α,γ-diaminobutyric acid in the molecular structure. PL-B was covalently immobilized on the surface of the fibers through the chemical reaction between the primary amino groups of PL-B and an active chlorine atom of the functional groups (fig. 2) [2].

Interaction between PL-B and endotoxin is based on ionic and hydrophobic forces. Ionic forces are through negatively charged phosphate groups in the lipid A portion of endotoxin and positively charged α,γ-diaminobutyric acid residue in PL-B. As the primary amino groups are also utilized for the immobilization

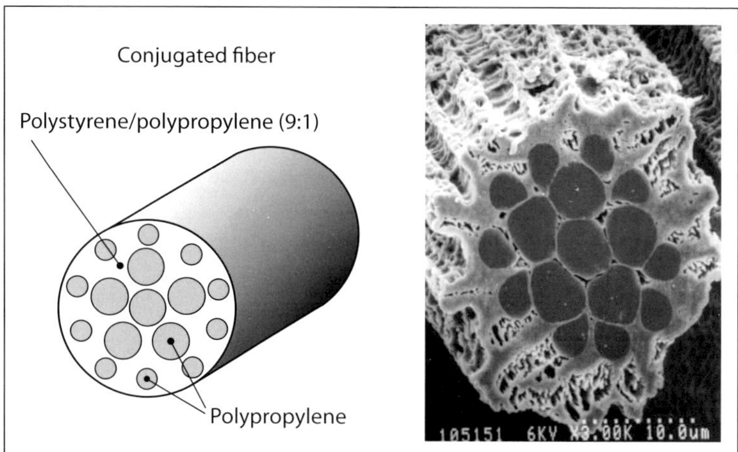

Fig. 2. The illustration on the left illustrates the schema of the cross-section of island-sea type-conjugated fiber filament. The island component is made of polypropylene polymer and the sea component is made of a mixed polymer of polystyrene and polypropylene (9:1). The picture on the right shows an electron micrograph of the cross-section of a fiber filament.

reaction, optimization of PL-B immobilization so as not to reduce the interaction forces was needed. It was found that the higher the number of primary amino groups in the fixed PL-B, the more the endotoxin removal capacity of PL-B-immobilized fiber was increased. PL-B was fixed so that there were 3–4 active primary amino groups in the fixed PL-B molecule.

An adsorbent compartment was built up with a knitted fabric made of PL-B-immobilized fiber (fig. 3). The PL-B-fixed knitted fabric was wrapped around a central pipe which had many side holes. By embedding this compartment into a case, the PMX cartridge was produced. Adopting this structural compartment, the homogeneous blood flow within the cartridge could be obtained. Blood enters the cartridge from the inlet and moves up within the central pipe. After going up the pipe, the blood goes out of the side holes in the central pipe crossing the layer of the fabric in touch with the adsorbent. By the development of a radial flow system within the cartridge, a homogeneous blood flow distribution could be achieved, contributing to the efficient utilization of the adsorbent.

Safety Evaluation

A healthy dog safety study was conducted to evaluate biocompatibility and safety of the PMX cartridge. Study subjects underwent a 2-hour extracorporeal

Fig. 3. An adsorbent compartment of a Toraymyxin cartridge.

treatment with a PMX cartridge. They were monitored for mean arterial pressure, hematocrit, platelet count, leukocyte count, serum proteins and transaminases for up to 3 h after treatment. All measurements remained or returned to acceptable ranges during and after treatment, and no study subjects died. No unfavorable (adverse) events were reported in this study.

Performances of PL-B-Immobilized Fiber Cartridge

In order to asses PMX performance in LPS removal, we have to introduce two concepts. The first is the 'endotoxin burden', which is the quantity of circulating endotoxins in a patient affected by (endotoxemic) septic shock. Based on the results from the MEDIC trial [3], high endotoxin activity level can be considered to be an endotoxin activity assay (EAA) ≥0.6; Monti et al. [4] showed EAA >0.9 in patients with septic shock. In turn, considering *Escherichia coli*-derived LPS, the corresponding LPS concentration is 7.0 ng/ml or 70 EU/ml (EU = endotoxin units); therefore, under the condition mentioned, the endotoxin burden of a 70-kg male is: 70 × 5,000 (total blood volume 5 l) = 350,000 EU.

The second concept is the 'device adsorbtion capability' (DAC), which looks at how many EUs a single PMX is able to remove from blood. The DAC of PMX is not reported officially. It was, however, possible to calculate it in a quite precise way from some previously published in vitro studies. It could also be estimated on the basis of the experience gained in the last years from the in vivo treatments monitored by EAA. The following table 1 shows the calculated values of the DAC of PMX (type: PMX-20R) in three different situations: (1) in vitro using solutions of hemoglobin, (2) in vitro with bovine blood (2 liters of recirculating blood) and (3) on a hypothetical septic patient with an infection from *E. coli* and an initial EAA value of 0.85 (which corresponds to an LPS concentration of 43 EU/ml for the *E. coli*). After a treatment with PMX, the hypothetical patient would have an EAA of 0.6 (which corresponds, to an LPS concentration of 15 EU/ml for *E. coli*). This hypothetical patient is absolutely realistic and consistent with data from different researchers (table 1) [4, 5].

How It Works

PMX is designed so that whole blood might be circulated during a hemoperfusion procedure. A large surface area could be obtained using fibrous material which has a small diameter. Furthermore, fibers have a porous surface structure resulting from the chemical reaction used to introduce a functional group. This induces an enlargement of the surface area of the adsorbents. On the other hand, pressure loss between the inlet and outlet of the cartridge remains low due to the use of a fibrous material, a knitted fabric as an adsorbent structure and the controlled homogeneous blood flow within a cartridge enabling stable whole blood hemoperfusion. The endotoxin DAC of PMX was evaluated in vitro comparing the circulation of 1.5 liters of calf serum solution containing purified *E. coli* LPS at the concentration of 10 ng/ml through the PMX cartridge at a flow rate of 100 ml/min and through a carrier fiber cartridge without PL-B. After 2 h of circulation, LPS concentration changes of the two solutions were evaluated. Only PMX could decrease the level of LPS.

Recently, Nishibori et al. [6] investigated the cellular components in the PMX column after hemoperfusion in septic patients. It is well known that different populations of leukocytes are activated during septic shock and change their adhesive phenotype. They therefore hypothesized that some populations

Table 1. Endotoxin adsorption capability device with in vitro and in vivo conditions

Toraymyxin LPS adsorption	Operating conditions	Initial endotoxin concentration	Adsorption capacity device
In vitro data (Tani et al. [9], 1992), static model Hb solution – PMX fiber only	solution of hemoglobin +2 g of PMX fibers; 1 h incubating time	686 EU/ml	1,404,480 EU
In vitro data (Sakai et al. [10], 1993), perfusion model bovine blood	in vitro perfusion – bovine blood: 2 liters; perfusion time: 2 h	400 EU/ml	640,000 EU
Septic shock patients with *E. coli* infection and EAA monitoring	hypothetical patient with *E. coli* infection; EAA = 0.9 before, ed EAA = 0.6 after	70 EU/ml	275,000 EU

Disregarding the first in vitro test (very far from the clinical setting), 1 PMX device is able to trap an average of around 300,000 EU in a standard session. This value is truly interesting because it is of the same order of magnitude of the 'endotoxin burden'.

of leukocytes might be adsorbed in the PMX column and removed from the blood circulation after the treatment. They demonstrated that PMX also specifically bound monocytes from the peripheral blood leukocytes of septic patients by means of an analysis of bound cells using immunocytochemical and electron microscopic techniques.

The specific removal of monocytes from septic patients may produce beneficial effects by reducing the interaction between monocytes and functionally associated cells, including vascular endothelial cells. The mechanism by which monocytes are specifically adsorbed on the surface of PL-B-immobilized fiber is still not clear, with the main hypotheses being physical characteristics of fibrous material of PL-B-immobilized fiber or some interactions mediated by endotoxin removal. This second appealing hypothesis (still to be demonstrated) is based on the assumption that binding sites of endotoxin and PL-B are different from those of endotoxin and monocytes. Therefore, PMX could remove monocytes by removing the LPS-monocyte complex. Anyhow, it is very important to remove endotoxin as a trigger of inflammatory reaction in the body. The fact that PMX can remove some activated immune cells is very interesting from the standpoint of its possible immunomodulatory effect for the treatment of severe sepsis and septic shock.

Advantages of Uses of Adsorbent Technology to Remove Endotoxin

A possible mechanism for blood purification techniques is to remove inflammatory mediators from the bloodstream which are involved in the pathogenesis and/or pathophysiology of severe sepsis and septic shock. There are multiple mediators that play an important role in the systemic inflammatory process such as endotoxin and exotoxin from exogenous sources or endogenous mediators such as cytokines, chemokines and various forms of lipids. Their molecular weight range is widely distributed, which is important when considering a 10- to 20-kd cutoff for a hollow fiber filter. Therefore, there is a need to choose appropriate modalities for selective mediator removal. The physicochemical aspect of mass transfer is the principle based on diffusion, filtration and adsorption. Renal replacement therapy, such as hemodialysis, hemofiltration and hemodiafiltration, uses the principles of diffusion and filtration, while the adsorption process is not affected by the cutoff point of hollow fiber. As a result, targeting molecular species becomes possible.

The basic molecular structure of endotoxin consists of a polysaccharide portion which includes an O-specific side chain, an inner and outer core region, and lipid A moiety which is an active center of LPS. Endotoxemia can occur in Gram-negative sepsis, or even Gram-positive sepsis, through bacterial or endotoxic translocation from the gut. The molecular aspect of endotoxin in blood or plasma is not clear. However, blood endotoxin is not always an LPS molecule itself, but can be in a variety of forms, such as membrane fragments, blebs and vesicles, in combination with bacterial phospolipids. In blood circulation, endotoxin may be bound by a large number of plasma components. In addition, LPS forms micelles in the blood. Therefore, the molecular weight is so big that it is difficult to remove through hemodialysis and the hemofiltration technique. Plasma exchange is one of the options for eliminating blood endotoxin. However, a lot of expensive fresh frozen plasma is not efficient and it carries some risk of infection as well.

Conclusion

Endotoxin has been discussed for a long time as the therapeutic target for the treatment of sepsis. However, reagents such as antiendotoxin monoclonal antibody and BPI (bactericidal/permeability increasing protein), which neutralizes endotoxin, did not improve the outcome of sepsis. This raised the question whether endotoxin is a real target for the development of novel therapeutics. Recently, favorable results of E5564 have been reported [7]. However, treatment with a phospholipid emulsion did not reduce 28-day all-cause mortality or reduce the onset of new organ failure in patients with suspected or confirmed Gram-negative severe sepsis [8]. It seems that endotoxin as a therapeutic target has mixed success and is still controversial.

The reason why some antiendotoxin reagents do not work well is not always clear. It may be related to the low efficiency of the reagent itself in neutralizing endotoxin. In the past a major question was to identify the more effective antiendotoxin strategy: endotoxin neutralization with reagents in the body or endotoxin extracorporeal removal from the blood. It seems possible to say after 15 years of clinical application that extracorporeal endotoxin removal by polymyxin B hemoperfusion is rational for the treatment of severe sepsis and septic shock. In addition, it is rational to use adsorbent technology to remove endotoxin from the blood as endotoxin is estimated to exist in the blood as a large molecule or aggregate.

It is very important to design an adsorbent which can exert a strong affinity with endotoxin in the blood. It may be possible to design an adsorbent using a different kind of ligand. However, PL-B is the most potent molecule that has a strong affinity with endotoxin and could construct a selective adsorbent which could work in the blood binding LPS even if linked to a multicomponent structure. PMX has been safely applied clinically for the treatment of severe sepsis and septic shock since 1994 in Japan and since 2002 in Italy, and is becoming more widely used year by year.

Another important therapeutic point of view is in relation to diagnostics. It is questionable whether blood endotoxin has been correctly measured in the previous antiendotoxin studies. Blood endotoxin measurement is still controversial. However, the newly developed EAA, which is an FDA-approved novel assay for endotoxin, is commercially available. This could enable endotoxin-targeted treatment by selecting an adequate patient population. PMX combined with EAA therapy is one of the more promising approaches for treating severe sepsis septic shock.

References

1 Bhor VM, Thomas CJ, Surolia N, Surolia A: Polymyxin B: an ode to an old antidote for endotoxic shock. Mol Biosyst 2005;1: 213–222.

2 Shoji H: Extracorporeal endotoxin removal for the treatment of sepsis: endotoxin adsorption cartridge (Toraymyxin). Ther Aher Dial 2003;7:108–114.

3 Marshall JC, Foster D, Vincent JL, Cook DJ, Dellinger RP, Opal S, Abraham E, Brett S, Smith T, Mehta S, Derzko A, Romaschin A: Diagnostics and prognostic implications of endotoxemia in critically illness: results of the MEDIC study. J Infect Dis 2004;190: 527–534.

4 Monti G, Terzi V, Mininni M, Colombo S, Vesconi S, Casella G: Polymyxin B hemoperfusion in high endotoxin activity level septic shock patients. Critical Care 2008;12(Suppl 2):P458, DOI: 10.1186/cc6679.

5 Novelli G, Rossi M, Poli L, Ferretti G, Ruberto F, Spoletini G, Levi Sandri GB, Mennini G, Morabito V, Berloco PB (Roma): Valutazione precoce dell'endotossinemia in pazienti affetti da sepsi nel periodo post operatorio mediante lo spectral's EAA (TM) endotoxin activity assay XXXIII Congresso nazionale SITO, Milano 13–15 Dicembre 2009.

6 Nishibori M, Takahashi HK, Katayama H, Mori S, Saito S, Iwagaki H, Tanaka N, Morita K, Ohtsuka A: Specific removal of monocytes from peripheral blood of septic patients by polymyxin B-immobilized filter column. Acta Med Okayama 2009;63:65–69.
7 Tidswell M, Tilllis W, LaRosa SP, Lynn M, Wittek AE, Kao R, Wheeler J, Gogate J, Opal SM, the Eritoran Sepsis Study Group: Phase 2 trial of Eritoran tetrasodium (E5564), a Toll-like receptor 4 antagonist, in patients with severe sepsis. Crit Care Med 2010;38:72–83.
8 Dellinger RP, Tomakyo JF, Angus DC, Opal S, Cupo MA, McDermont S, Ducher A, Calandra T, Cohen J, the Lipid Infusion and Patient Outcomes in Sepsis (LIPOS) Investigators: Efficacy and safety of a phospholipid emulsion (GR270773) in Gram-negative severe sepsis: results of a phase II multicenter, randomized, placebo-controlled, dose-finding clinical trial. Crit Care Med 2009;37:2929–2938.
9 Tani T, Chang TMS, Kodama M, Tsuchiya M: Endotoxin removed from hemoglobin solution using polymyxin B immobilized fiber followed by a new turbidometric endotoxin assay. Biomat Art Cells & Immob Biotech 1992;20:457–462.
10 Sakay Y, Shoji H, Kobayashi T, Terada R, Sugaya H, Murakami M, Moriyama K, Minaga M, Kunimoto T, Takeyama T: New extracorporeal blood purification devices for critical care medicine under development. Therapeutic Plasmapheresis 1993;12: 837–842.

Tohru Tani, MD, PhD
Department of Surgery, Shiga University of Medical Science
Tsukiwa-cho, Seta
Otsu City, Shiga, 520-2121 (Japan)
Tel. +81 077 548 2237, Fax +81 077 548 2240, E-Mail tan@belle.shiga-med.ac.jp

Ronco C, Piccinni P, Rosner MH (eds): Endotoxemia and Endotoxin Shock: Disease, Diagnosis and Therapy. Contrib Nephrol. Basel, Karger, 2010, vol 167, pp 45–54

Mechanisms of Polymyxin B Endotoxin Removal from Extracorporeal Blood Flow: Molecular Interactions

S. Vesentini · M. Soncini · G.B. Fiore · A. Redaelli

Department of Bioengineering, Politecnico di Milano, Milan, Italy

Abstract

The outer leaflet of Gram-negative bacteria membrane contains a great amount of lipopolysaccharides, also known as endotoxins, which play a central role in the pathogenesis of septic shock. It has been demonstrated that the polymyxin B (PMB) molecule has both antibacterial and antiendotoxin capabilities; in fact, it is able to compromise the bacterial outer membrane and bind lipopolysaccharides, thereby neutralizing its toxic effects. Extracorporeal hemoperfusion treatments based on cartridges containing PMB-immobilized fibers (Toraymyxin PMX-F; Toray Industries, Tokyo, Japan) are used to remove endotoxins circulating in the blood flow. In this study, we focused on the characterization of the interactions occurring in the formation of the PMB-endotoxin complex at the molecular level. In particular, the molecular mechanics approach was used to evaluate the interaction energy and eventually the interaction force between the two molecules. PMB was faced with five molecular portions of lipopolysaccharides differing in their structure. The interaction energy occurring for each molecular complex was calculated at different intermolecular distances and the binding forces were estimated by fitting interaction energy data. Results show that the short-range interactions between PMB and endotoxins are mediated mainly by hydrophobic forces, while in the long term, the complex formation is driven by ionic forces only. Maximum binding forces calculated via molecular mechanics for the PMB-endotoxin complex are in the range of 1.39–3.79 nN. Understanding the interaction mechanism of the single molecular complex is useful both in order to figure out the molecular features of such interaction and to perform higher scale level analysis, where such nanoscale detail is impractical but could be used to account for molecular behavior at a coarse level of discretization.

Sepsis is a generalized infection of an organism with presence of bacteria in the blood flow (bacteremia). Recently, in parallel to clinical and research developments concerning antibiotic treatments, antiendotoxinic therapies have been explored in order to define new approaches able to act directly on endotoxins, which are the main constituent of the outer leaflet of the Gram-negative bacteria external wall.

An interesting antiendotoxin strategy is represented by the use of an extracorporeal device (Toraymyxin PMX-F; Toray Industries, Tokyo, Japan), wherein the polymyxin B (PMB) antibiotic is grafted on sorbent material and is able to provide selective removal of circulating endotoxins. PMB is a cationic amphiphilic cyclic decapeptide able to compromise the bacterial membrane (antibacterial effect) and bind the residual endotoxins (antiendotoxin effect) [1]. The use of an extracorporeal cartridge avoids the negative effects produced by circulating PMB molecules, which have been demonstrated to be nephrotoxic and neurotoxic when released at the local and systemic level. In fact, in the Toraymyxin device, PMB molecules are covalently grafted on the fiber surface (a knitted tissue of polypropylene and α-chloroacetamide-methylpolystyrene) packed inside the cartridge.

The highly amphiphilic character of PMB is due to the presence of hydrophobic groups alternating with positively charged hydrophilic diaminobutyric acid (Dab) residues [2]. Such a structure allows PMB molecules to bind to a specific region of the lipopolysaccharides (LPS), the lipid-A region, and to neutralize LPS toxic effects by acting on the spatial organization of the LPS fatty acid chains.

LPS consist of a hydrophilic polysaccharide domain bound to a hydrophobic lipid tail (lipid A), which is the LPS portion embedded in the bacterial outer membrane (fig. 1). The molecular structure of the lipid A is constituted by a diphosphorylated D-glucosamine disaccharide backbone acylated by up to seven asymmetric fatty acid chains. The polysaccharide region is made by a sequence of oligosaccharide units (O-antigen) and several rare sugars (inner and outer core). The O-antigen structure provides serotype specificity and polar properties to the overall LPS structure, while lipid A is responsible for most of the pathological effects. The O-antigen structure varies markedly among different bacterial species, while the inner and outer core portions show slight interbacterial variability, and the lipid-A portion is the most conserved structure in LPS [3–5].

The expression of endotoxic activity is regulated by a specific equilibrium between hydrophilic (diphosphorylated disaccharide backbone and ionization state of phosphate groups) and hydrophobic (number, length and position of fatty acid chains) regions, which drive the molecular mechanisms underlying the adverse biochemical effects induced by LPS binding with specific immune system receptors of healthy cells in vivo. The active state conformation is promoted by the 2-keto-3-deoxy-manno-octonoic acid dimer (Kdo) of the inner core directly connected to the lipid-A saccharide backbone [6, 7]. In the Kdo-Kdo-lipid-A complex, named ReLPS (fig. 1a), the Kdo's negative charges play a regulatory effect on the ionization state of the disaccharide backbone, which

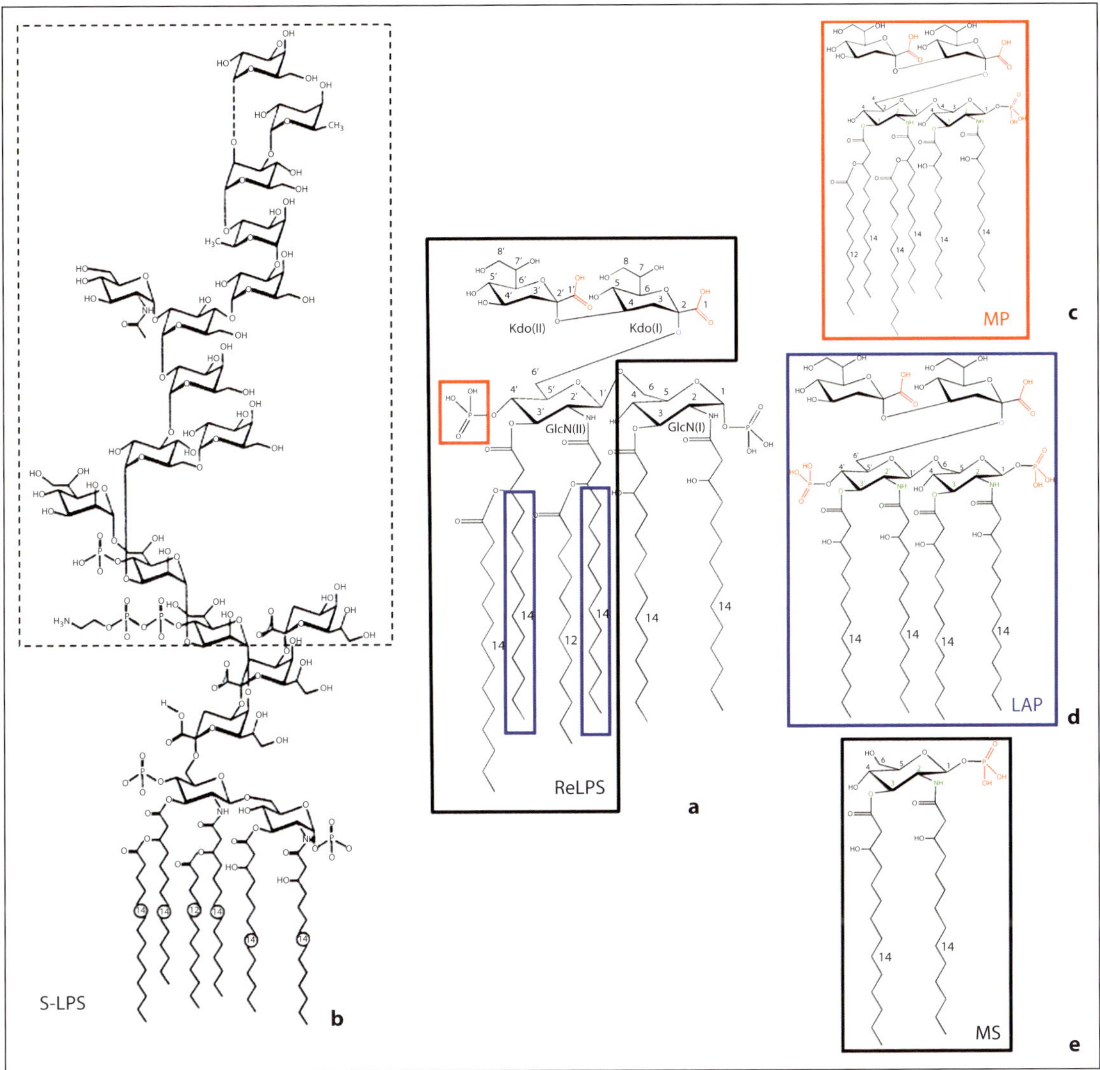

Fig. 1. Molecular structure of LPS. **a** The minimum structure required for accomplishing the endotoxin activity of LPS (ReLPS), which consists of lipid A (consisting of a D-glucosamine disaccharide – GlcN(I) and GlcN(II) – with 2 phosphate groups attached to positions 1 and 4′, and by 6 fatty acid chains linked to the saccharide backbone) and the inner-core region (composed by the α-(2, 4)-dimer of 2-keto-3deoxy-manno-octonoic acid, named the Kdo-Kdo group). **b** Complete structure of the LPS molecule (S-LPS) obtained adding the O-antigen and the outer core (marked with dotted contour) to the inner core and the lipid-A structures of the ReLPS. **c** Structure of ReLPS monophosphate (MP), obtained removing the phosphate group at the position 4 from the ReLPS structure (the removed phosphate group is marked in red in panel **a**). **d** Precursor of lipid A (LAP), derived from ReLPS model without fatty acid side secondary chains (the removed fatty acid chains are marked in blue in panel **a**). **e** ReLPS monosaccharide (MS), without one of the sugars of its saccharide backbone [GlcN (II)] and related fatty acid chains, and the Kdo-Kdo group (the removed structures are marked by means of a black contour in panel **a**). Numbers close to the fatty acid chains indicate the number of carbon atoms.

influences the fatty acid chain configuration. The amphiphilic character of PMB structure makes PMB able to selectively bind the lipid-A region of LPS, thus neutralizing their adverse effects on cells.

The present paper is a summary of a more extended study [8] that investigated the interaction forces generated at the molecular level within the LPS-PMB complexes and clarified the mechanisms of complex formation. In particular, different molecular models of LPS structures were constructed and the interaction forces between LPS and PBM were estimated at different distances using the molecular mechanics approach.

Methods

The molecular model of PMB together with five different LPS structures were developed and evaluated. The molecular analyses were carried out by means of the computational chemistry commercial package Hyperchem® 6.01 (Hypercube, Canada) using a force field specific for biological molecules (Amber 3) and simulating an implicit water-like environment by means of a dielectric constant equal to 78.

The PMB Model

The active structure of PMB was modeled on the basis of the atomic coordinates by Pristovsek and Kidric [2]. In the active state, the PMB molecule is in the typical conformation bound to the lipid-A region of endotoxins. The Gasteiger and Marsili empiric method [9] was used to determine the atomic charges for the PMB model. A $+0.5e$ total charge (e is the electron charge) was assigned to the amine group (NH_2) of each Dab residue, equally shared between the two hydrogen atoms. The molecular structure was not optimized in a water-like environment, with the aim of preserving the bound PBM active structure [2]. In this way, the highly amphiphilic conformation of the PMB model was maintained with the hydrophobic residues lying on one side and the hydrophilic residues lying on the opposite side. The total length of the molecule was 2.6 nm.

The LPS Models

Five different LPS structures were considered to be faced with PMB (fig. 1). The complete endotoxin structure (S-LPS) is shown in figure 1b and the minimum structure required for accomplishing the endotoxin activity of LPS, which is the Kdo-Kdo-lipid-A complex (ReLPS), is shown in figure. 1a. The ReLPS and S-LPS models were obtained according to the atomic models proposed by Kastowsky et al. [6]. In addition, in order to evaluate the effects of ionic and hydrophobic interactions driving the molecular complex formation, three other structures were obtained by removing some functional groups from the ReLPS structure. In particular, the monophosphate precursor (MP) was considered to reduce the ionic interaction by eliminating the phosphate group in position 4 (fig. 1c), the lipid-A precursor (LAP) was obtained from ReLPS to reduce the hydrophobic interaction by removing the secondary fatty acid chains (fig. 1d), and the monosaccharide (MS) was modeled to verify the effects of the decrease of both ionic and hydrophobic interactions caused by the lack of an entire sugar from its saccharide backbone and the corresponding phosphate group and fatty acid chains (fig. 1e).

The Gasteiger and Marsili method [9] based on partial equalization of the orbital electronegativity scheme was applied to calculate the atomic charges. The ionization state of the ReLPS molecular model was assigned according to Din et al. [10]. Negative charges are located on the two phosphate groups of lipid A (–1*e* for each phosphate group) and on the two carboxylate groups of Kdo (–0.5*e* charge was assigned for the two oxygens of each carboxylate group). The S-LPS ionization state was set according to atomic charges used for the Kdo-Kdo-lipid-A portion in the ReLPS model; the third Kdo residue was set accordingly to the other two Kdo groups.

Energetic optimization was performed for both S-LPS and ReLPS models using the steepest descent followed by the Polak-Ribiere minimization algorithm in order to obtain the initial structure corresponding to the minimum potential energy of the molecular system. The combination of these two minimization algorithms allows one to obtain an adequate optimization procedure saving computational costs.

The MP, LAP and MS models were derived from the optimized ReLPS model, maintaining the ionization state; the precursor models were optimized following the same minimization scheme used for large complexes.

LPS-PMB Interaction Force Calculation

The five endotoxin models (S-LPS, ReLPS, MP, LAP, MS) were faced with PMB molecular model in order to obtain five different molecular complexes ($C_{S\text{-}LPS}$, C_{ReLPS}, C_{MP}, C_{LAP}, C_{MS}). Each LPS-PMB molecular complex was energetically minimized to obtain a stable coupling between PMB and LPS.

The optimized complexes were then used as initial configurations in order to estimate the interactions at different intermolecular distances. Intermolecular distance was imposed to the molecular system moving the LPS molecule with respect to the PMB position and energy optimization was performed in two steps to achieve the minimum energy configuration. Specifically, in the first step the PMB structure is fixed and LPS is free to move; in the second step both molecules can move. The single reaction coordinate r_{PL} was imposed by acting on the LPS center of mass (CM_L) and maintaining the position of the PMB center of mass (CM_P) fixed; the LPS was moved along the line connecting CM_L and CM_P, applying step distance increments of 0.05 nm. Specifically, r_{PL} represents the intermolecular distance calculated as:

$$r_{PL} = \sqrt{(x_{CM_P} - x_{CM_L})^2 + (y_{CM_P} - y_{CM_L})^2 + (z_{CM_P} - z_{CM_L})^2} \tag{1}$$

$$x_{CM} = \frac{\sum_k m_k x_k}{\sum_k m_k}; \qquad y_{CM} = \frac{\sum_k m_k y_k}{\sum_k m_k}; \qquad z_{CM} = \frac{\sum_k m_k z_k}{\sum_k m_k} \tag{2}$$

where x_k, y_k, z_k are the atomic spatial coordinates and m_k is the atomic weight of each kth atom.

At each applied intermolecular distance, the interaction energy (E'_{PL}) was calculated by subtracting the potential energy of the PMB (E_P) and LPS (E_L) to the potential energy of the overall molecular system (E_{TOT}):

$$E'_{PL} = E_{TOT} - E_P - E_L \tag{3}$$

The interaction energy (E_{PL}) between the two molecules as a function of the intermolecular distance (r_{PL}) was obtained interpolating E'_{PL} versus r_{PL} values by means of a Lennard-Jones (L-J) potential:

$$E_{PL} = 4\varepsilon\left[\left(\frac{\sigma}{r_{PL}}\right)^{12} - \left(\frac{\sigma}{r_{PL}}\right)^{6}\right] \tag{4}$$

where ε and σ are the L-J parameters determined through a best-fit algorithm. In particular, ε represents the equilibrium energy ($\varepsilon = E_{PLmin}$) and σ the equilibrium length ($\sigma 2^{1/6} = r_{PLmin}$).

The binding force is calculated as the first order derivative of E_{PL} with respect to the reaction coordinate r_{PL}:

$$F(r_{PL}) = -\frac{dE_{PL}}{dr_{PL}} = 24\varepsilon\left[2\frac{\sigma^{12}}{r_{PL}^{\;13}} - \frac{\sigma^{6}}{r_{PL}^{\;7}}\right] \tag{5}$$

Results and Discussion

Following the interaction calculation procedure, the interaction energy was evaluated for all complexes in order to quantify the L-J interaction energy curve characterizing each PMB-LPS complex and the binding forces were derived from L-J-interpolated curves as a function of the intermolecular distance.

For the LPS-PMB complexes, the major contribution to the interaction energy is due to the hydrophobic effects, being the van der Waals energy term equal to 96% of the total interaction energy. The only exception is represented by the C_{MS} complex where the van der Waals contribution is 44% of the total energy due to the marked reduction of the number of fatty chains in this LPS precursor.

The range of maximum force values calculated for each complex is reported in table 1 (column 1). Large complexes show higher binding forces than small complexes (table 1, column 1). This difference is definitely due to the suppression of functional groups in the small LPS precursors, and is particularly marked for C_{LAP} and C_{MS}, but negligible for C_{MP}. The observed behavior is likely due to the number of fatty acid chains within LPS models [11]. In fact, in the S-LPS, ReLPS and MP models, the lipid-A portion consists of 6 fatty acid chains, while in the LAP and MS it has 4 and 2 fatty acid chains, respectively. The highest binding interactions are measured for those complexes containing the complete structure of lipid A ($C_{S\text{-}LPS}$ and C_{ReLPS}) according to Din et al. [10]. The maximum binding force calculated for $C_{S\text{-}LPS}$ and C_{ReLPS} complexes ranges between 1.94 and 3.79 nN. Lower binding forces (range 1.39–1.85 nN) are observed for the other three complexes (C_{MP}, C_{LAP} and C_{MS}).

In order to elucidate the binding mechanisms occurring between PMB and LPS structures, figure 2 shows a sequence of snapshots at different interaction distances for the ReLPS-PMB complex. For each simulated configuration, the

Table 1. Characterization of the LPS-PMB complexes: binding force and L-J curve parameters

LPS-PMB complex	F_{max} nN	ε kJ/mol	σ nm
$C_{S\text{-}LPS}$	2.69±1.10	274.0±33.0	0.63±0.13
C_{ReLPS}	1.13±1.04	196.0±35.0	0.58±0.16
C_{MP}	1.28±0.57	222.5±10.7	0.41±0.14
C_{LAP}	0.98±0.41	156.3±21.8	0.72±0.21
C_{MS}	0.96±0.84	132.9±31.7	0.46±0.10

F_{max} = Maximum binding force; ε = equilibrium energy; σ = equilibrium length.

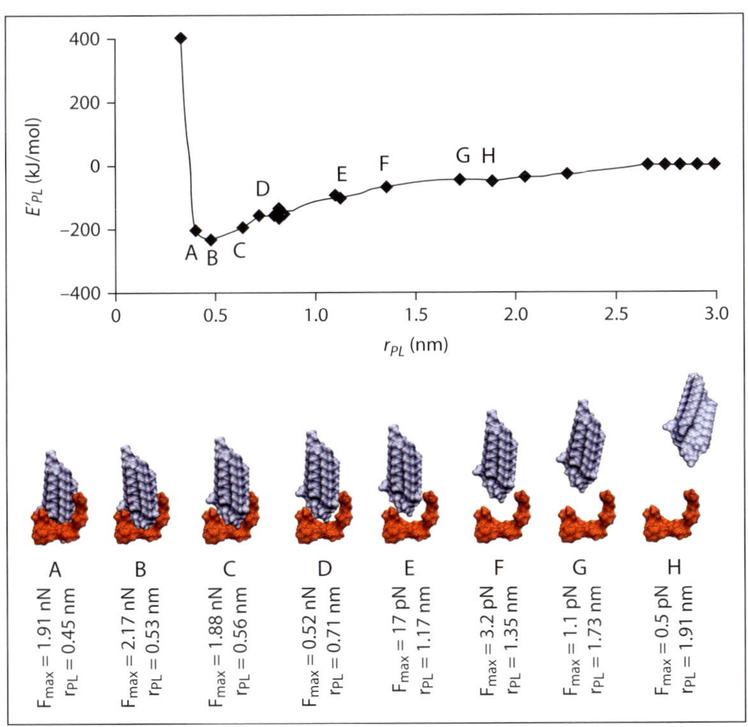

Fig. 2. L-J interaction energy curve obtained for the molecular complex C_{ReLPS} (upper graph) and related sequence of configurations explored at different intermolecular distance (lower panel). Short-range configurations: snapshot A, B and C are characterized by low intermolecular distances and high binding forces (both ionic and hydrophobic interactions are involved at this stage of the complex formation). Long-range configurations: snapshots D–H are characterized by smaller binding forces (hydrophobic effects drive the complex formation).

intermolecular distance and the maximum binding force calculated using the L-J parameters obtained by fitting the L-J energy curve ($\sigma = 0.74$ nm and $\varepsilon = 231$ kJ/mol for C_{ReLPS}) are also reported. In figure 2, the short-range (panels A, B and C) and the long-range (panels D–H) conformations are shown, representing different configurations of the complex at increasing intermolecular distances. In the short-range interaction, the binding force is about 2 nN and both ionic and hydrophobic effects are involved in the binding mechanisms. In this phase the fatty acid chains are responsible for the stabilization of the complex. In long range interaction, the binding force decreases by orders of magnitude when the distance between molecules is progressively increased. In this second phase, the ionic forces become prevalent.

The results obtained via molecular mechanics simulations for the binding force generated in the LPS-PMB complexes are slightly higher than, but of the same order of magnitude as, the binding force values obtained by means of atomic force microscopy for different biological complexes: (1) the antigen-antibody complex (0.1 0.05 nN; [12]), (2) the protein complex involved in cell adhesion (0.4 nN; [13]) and (3) the complex formed by the receptor protein of *Escherichia coli* bacterium with its proteic substrate (0.4–0.9 nN; [14]). Comparing the results obtained in this work for the LPS-PMB complexes with data from the literature about other common biological complexes, we can hypothesize that both large (S-LPS, ReLPS) and small precursor structures (MP, LAP and MS) are able to establish very stable complexes with PMB, proving the antiendotoxic ability of PMB molecules.

Concluding Remarks

The effectiveness of antiendotoxin treatments is based on the stable and specific interaction occurring between endotoxin and PMB molecules at the molecular level. With the aim of characterizing the stability of PMB-LPS binding interaction, a molecular model of PMB was interfaced with five molecular models of LPS differing in their structure, and molecular mechanics simulations were performed at different intermolecular distances in order to calculate the interaction energies of the complex. Binding forces were calculated by fitting interaction energy data with a L-J energy function. Calculated maximum binding forces for different PMB-LPS complexes are in a range between 1.39 and 3.79 nN (maximum values in table 1), which is comparable with data obtained for other complexes of biological relevance and proves that PMB is able to provide stable binding with large LPS structures and their precursors. The interaction mechanism is driven by ionic and hydrophobic forces at the short range, while ionic forces dominate the complex formation in the long-range interactions.

The characterization of LPB-PMB complexes at the molecular level can be used to set up dynamic simulations at a higher-scale level, applying a multiscale approach in order to evaluate the efficacy of a device for endotoxin removal. In particular, at the microscale level, the fluid-dynamic field within the cartridge can be taken into account. Hence, the competition between the drag forces exerted on LPS by the fluid flow and the binding forces exerted by grafted PMB molecules can be investigated [see the chapter by Fiore et al., pp. 55–64].

Acknowledgments

The authors thank Dr. P. Pristovsek of the National Institute of Chemistry, Ljubljana, Slovenia, for providing the atomic coordinates of PMB molecule and Dr. M.J. Kastowsky of the Max Planck Institute, Jena, Germany, for providing the atomic coordinates of S-LPS and ReLPS lipopolysaccharides.

References

1 Morrison DC, Jacobs DM: Binding of polymyxin B to the lipid A portion of bacterial lipopolysaccharides. Immunochemistry 1976;13:813–818.

2 Pristovsek P, Kidric J: Solution structure of polymyxins B and E and effect of binding to lipopolysaccharide: an NMR and molecular modeling study. J Med Chem 1999;42:4604–4613.

3 Brade H, Brade L, Rietschel ET: Structure-activity relationships of bacterial lipopolysaccharides (endotoxins). Current and future aspects. Zentralbl Bakteriol Mikrobiol Hyg [A] 1988;268:151–179.

4 Rietschel ET, Brade L, Schade U, Seydel U, Zäringer U, Kusumoto S, Brade H: Bacterial endotoxins: properties and structure of biologically active domains; in Schrinner E, Richmond MH, Seibert G, Schwarz U (eds): Surface Structures of Microorganisms and Their Interactions with the Mammalian Host. Weinheim, VCH, 1993, pp 1–41.

5 Zahringer U, Lindner B, Rietschel ET: Molecular structure of lipid A, the endotoxic center of bacterial lipopolysaccharides. Adv Carbohydr Chem Biochem 1994;50:211–276.

6 Kastowsky M, Obst S, Bradaczek H: Molecular dynamics simulations of six different fully hydrated monomeric conformers of *Escherichia coli* re-lipopolysaccharide in the presence and absence of Ca^{2+}. Biophys J 1997;72:1031–1046.

7 Wang Y, Hollingsworth RI: An NMR spectroscopy and molecular mechanics study of the molecular basis for the supramolecular structure of lipopolysaccharides. Biochemistry 1996;35:5647–5654.

8 Vesentini S, Soncini M, Zaupa A, Silvestri V, Fiore GB, Redaelli A: Multi-scale analysis of the toraymyxin adsorption cartridge. Part I: molecular interaction of polymyxin B with endotoxins. Int J Artif Organs 2006;29:239–250.

9 Gasteiger J, Marsili M: Iterative partial equalization of orbital electronegativity – a rapid access to atomic charges. Tetrahedron 1980;36:3219–3222.

10 Din ZZ, Mukerjee P, Kastowsky M, Takayama K: Effect of pH on solubility and ionic state of lipopolysaccharide obtained from the deep rough mutant of *Escherichia coli*. Biochemistry 1993;32:4579–4586.

11 Yin N, Marshall RL, Matheson S, Savage PB: Synthesis of lipid A derivatives and their interactions with polymyxin B and polymyxin B nonapeptide. J Am Chem Soc 2003;125:2426–2435.

12 Dammer U, Hegner M, Anselmetti D, Wagner P, Dreier M, Huber W, Guntherodt HJ: Specific antigen/antibody interactions measured by force microscopy. Biophys J 1996;70:2437–2441.

13 Dammer U, Popescu O, Wagner P, Anselmetti D, Güntherodt HJ, Misevic NC: Binding strength between cell adhesion proteoglycans measured by atomic force microscopy. Science 1995;267:1173–1175.

14 Vinckier A, Gervasoni P, Zaugg F, Ziegler U, Lindner P, Groscurth P, Pluckthun A, Semenza G: Atomic force microscopy detects changes in the interaction forces between GroEL and substrate proteins. Biophys J 1998;74:3256–3263.

Simone Vesentini
Department of Bioengineering, Politecnico di Milano
Piazza Leonardo da Vinci, 32
IT–20132 Milano (Italy)
Tel. +39 02 23993375, Fax +39 02 23993360, E-Mail simone.vesentini@polimi.it

Ronco C, Piccinni P, Rosner MH (eds): Endotoxemia and Endotoxin Shock: Disease, Diagnosis and Therapy. Contrib Nephrol. Basel, Karger, 2010, vol 167, pp 55–64

Mechanisms of Polymyxin B Endotoxin Removal from Extracorporeal Blood Flow: Hydrodynamics of Sorption

G.B. Fiore · M. Soncini · S. Vesentini · A. Redaelli

Department of Bioengineering, Politecnico di Milano, Milan, Italy

Abstract

The removal of blood endotoxins with the Toraymyxin extracorporeal sorption device exploits the capability of immobilized polymyxin B (PMB) to bind endotoxins stably with a high specificity. Although adsorption is a molecular-scale mechanism, it involves hydrodynamic phenomena in the whole range from the macroscopic down to the supramolecular scales. In this paper we summarize our experience with a computational, multiscale investigation of this device's hydrodynamic functionality. 3D computational fluid dynamics models were developed for the upper-scale studies. The flow behavior in the sorbent material was either modeled as a homogeneous Darcy's flow (macroscale study), or described as the flow through realistic geometrical models of its knitted fibers (mesoscale study). In the microscale study, simplified 2D models were used to track the motion of modeled endotoxin particles subjected to the competition of flow drag and molecular attraction by the fiber-grafted PMB. The results at each scale level supplied worst-case input data for the subsequent study. The macroscale results supplied the peak velocity of the flow field that develops in the sorbent. This was used in the mesoscale analysis, yielding a realistic range for the shear stresses in the fluid next to the fiber surface. With wall shear stresses in this range, endotoxin particle tracking was studied both in the vicinity of a single immobilized PMB molecule, and in the presence of a layer of PMB molecules evenly distributed at the fiber surface. Results showed that the capability to seize endotoxin molecules extends at least at a distance of 10–20 nm from the surface, which is one order of magnitude greater than the stable intermolecular bond characteristic distance. We conclude that a multiscale approach has the power to provide a comprehensive understanding, shedding light both upon the physics involved at each scale level and the mutual interactions of phenomena occurring at different scales.

Sepsis has been reported to be the first cause of death in noncoronary ICUs [1], and the 11th cause of overall mortality [2] in the United States. Aged and/or immunodepressed critical-conditions patients are the subjects for whom sepsis represents a threatening cause of mortality in the ICU [3, 4]. In an attempt to act on the primary trigger of the inflammatory process (i.e. the endotoxins) [5], a specific protein, polymyxin B (PMB), was discovered in the 1970s and found to be effective against endotoxin-induced hemodynamic shock [6–8], but extremely toxic for the kidney and central nervous system [9–11]. However, exploiting the binding capability of immobilized PMB with respect to endotoxins by extracorporeal means was proposed as a way to avoid the drawbacks related to systemic toxicity. PMB molecules covalently grafted to an inert fiber are capable of acting as adsorption agents to seize endotoxins from an extracorporeal blood stream lapping the fiber surface.

Adsorption is a molecular-scale mechanism; however, its accomplishment involves hydrodynamic phenomena that take place in the whole scale range from the macroscopic level down to the supramolecular scale. Indeed, endotoxin removal efficiency must rely on conveying a blood stream through a small adsorption cartridge so as to ensure an extended and efficient interaction of the blood stream with the active fiber surface. This is obtained with a careful hydrodynamic design of the blood device. In 1994, Toray Medical Co. (Tokyo, Japan) presented the Toraymyxin extracorporeal device for endotoxin removal on the Japanese market. It was CE marked in 1998 and has been available in Europe since 2002. The patient's venous blood enters this cartridge through a cylindrical blood distributor. It flows radially through a fibrous material (which is rolled up the central distributor), then into a collection chamber and finally to the device outlet. The fibrous region contains immobilized PMB, which is meant to reduce the blood endotoxin content. The effectiveness of the treatment with the Toraymyxin device was demonstrated clinically [12–15], although questions have been raised about its core functionality [16–18].

This paper summarizes our experience with a multiscale fluid-dynamic analysis of the Toraymyxin cartridge [19]. We analyzed the functional mechanisms at three different scales. At the macroscale, computational fluid dynamics (CFD) was used to characterize the whole device in terms of its overall hydrodynamic variables. At the mesoscale, the effect of the knitted fiber structure was analyzed and CFD simulations were run with input fluid velocities equal to the worst-case data calculated at the macroscale level. At the microscale level, the capturing efficacy of immobilized PMB was investigated by balancing the forces exerted on endotoxin molecules by the fluid drag and molecular interactions, the latter being studied in detail in a parallel work dealing with the endotoxin/PMB molecular mechanics [see the chapter by Vesentini et al., pp. 45–54].

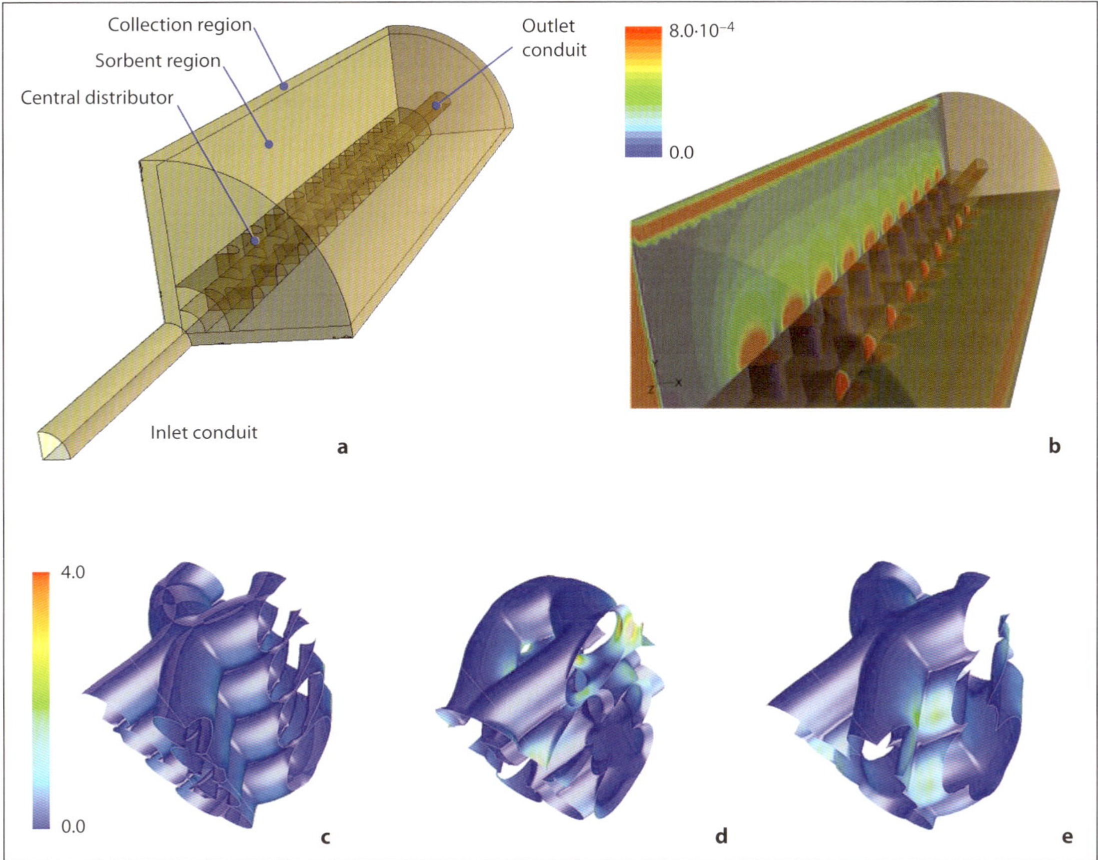

Fig. 1. **a** 3D sketch of the macroscale geometrical model of the Toraymyxin filter: one fourth of the overall geometry is modeled thanks to the device's symmetry. **b** Macroscale study: map of the fluid velocity in the sorbent region (m/s). Peak velocity in the sorbent takes place immediately beyond the impact surface to the sorbent itself. **c–e** Mesoscale study: WSS maps at the fiber walls (Pa). WSS is quite homogeneous for cluster A (**c**); clusters B and C (**d** and **e**, respectively) display uneven WSS patterns.

Materials and Methods

Macroscale Study: Global Hydrodynamics within the Device

Our CFD study of the whole device started from building a 3D geometrical model of the blood flow path (fig. 1a) from technical drawings provided by the manufacturer. Thanks to symmetries, only one fourth of the real shape was modeled, and meshed with 1,560,000 tetrahedral cells.

Macroscale CFD simulations were run with Fluent software (Ansys Inc.) under the following general assumptions: laminar flows, Newtonian, homogeneous and incompressible fluids.

The device was modeled to work at the standard conditions for the Toraymyxin cartridge, i.e. with a 100-ml/min inlet blood flow rate at 37°C. The sorbent material was hydraulically modeled as a permeable porous medium behaving according to Darcy's law: $\nabla p = -\frac{\mu}{\alpha}\mathbf{v}$, where ∇p is pressure gradient, $\mathbf{v}$ is the so-called 'apparent' fluid velocity vector, μ is fluid viscosity and α is the porous medium's permeability. Two values for α were considered ($5.90 \cdot 10^{11}$ and $7.05 \cdot 10^{11}$ m^{-2}) based on the manufacturer's characterization data. Blood density was set at ρ = 1,060 kg m^{-3} and viscosity values were varied in the range $\mu = (2.7–4.0) \cdot 10^{-3}$ kg m^{-1} s^{-1} [20].

Mesoscale Study: Dynamics of Flow through the Fibers

When modeling the hydrodynamic phenomena at characteristic dimensions in the sub-millimetric scale, it is no longer possible to ignore the real geometrical features of the sorbent material, which is made up with knitted fibers. A single-layer model was used in the present study as a building block to define multiple-layer clusters which represent the way how subsequent fiber layers randomly superimpose during manufacturing. Each cluster comprised 4 knitted layers in our model, with cluster A built with the fibers of different layers perfectly aligned, and clusters B and C built with differently translated fiber layers. For each cluster, the fluid volume (obtained by subtraction of the cluster volume from a solid volume) was meshed with approximately 230,000 tetrahedral cells.

The general assumptions for the mesoscale study were the same as for the macroscale study, with blood density ρ = 1,060 kg m^{-3} and viscosity $\mu = 4 \cdot 10^{-3}$ kg m^{-1} s^{-1}. CFD simulations were run with the inlet velocity set equal to the largest value obtained in the macroscale simulations in the sorbent region, and results were post-processed estimating the wall shear stress (WSS) range of values to be used in the subsequent microscale study.

Microscale Study: Competition between Dragging and Binding of Endotoxin Molecules

Submicrometric characteristic dimensions were considered in the microscale study. At this scale, it is possible to describe the motion of an endotoxin molecule as a result of the forces acting on it, namely the drag force due to fluid flow and the molecular interaction forces, as calculated with molecular mechanics methods. The scale of the analyzed phenomena is small enough to let one isolate the hydrodynamics of the sole plasma sublayer adjacent to the fiber surface. Therefore, the problem was reduced to a 2D Cartesian problem, with the z-axis lying on the fiber surface and oriented as the direction of the adjacent plasma flow, and the positive y-coordinates measuring the distance from the surface. In the fluid, a shear flow of plasma was considered, with velocity parallel to the z-axis and proportional to the y-coordinate with a slope depending on the shear rate. This velocity profile was assumed to be undisturbed in the presence of moving particles. Density ρ = 1,035 kg m^{-3} and viscosity $\mu = 1.6 \cdot 10^{-3}$ kg m^{-1} s^{-1} were used for plasma. Simulations were run with five different WSS values in the interesting range identified by the mesoscale results.

The suspended endotoxin molecules were represented as spherical particles, characterized by radii calculated on the basis of the respective van der Waals molecular volumes, and assumed to obey the Stokes' law when dragged by the flow:

$$\mathbf{F}_d = 6 \pi \mu r_{eq} (\mathbf{v} - \mathbf{v}_p)$$

where $\mathbf{F}_d$ is the drag force vector and $\mathbf{v} - \mathbf{v}_p$ is the difference between the fluid velocity vector and the particle velocity vector. The complete endotoxin structure (S-LPS, equivalent radius r_{eq} = 1.56 nm) and the minimum structure displaying an endotoxin activity (ReLPS, r_{eq} = 1.07 nm) were considered [21].

Grafted PMB molecules were placed on the fiber wall with their binding site facing outwards at y = 2.6 nm, in consideration of the overall molecule length. The dynamics of endotoxin particles was studied both in the vicinity of a single immobilized PMB molecule and in the presence of a layer of PMB molecules evenly distributed at the fiber wall. In the latter case, based on data supplied by the manufacturer, an average surface concentration of 0.25 molecules/nm^2 was considered, which implied (in our 2D domain) an average linear concentration of 1 molecule every 2 nm.

The molecular interaction force was applied to the endotoxin particle's center of mass, pointing towards the PMB's binding site. Its modulus was expressed as a function of the intermolecular distance r as a Lennard-Jones force (positive values are attractions):

$$F_{LJ} = -24\varepsilon\left[2\frac{\sigma^{12}}{r^{13}} - \frac{\sigma^{6}}{r^{7}}\right]$$

where σ and ε are the parameters of the corresponding expression for the interaction energy (σ = 0.63 nm and ε = 274 kJ/mol for the S-LPS-to-PMB interaction and σ = 0.58 nm and ε = 196 kJ/mol for the ReLPS-to-PMB interaction, chosen based on the smallest peak attraction forces; see our companion chapter).

Results

Macroscale Study

Simulation results show that the velocity distribution is quite uniform within the sorbent region: an example velocity map is displayed in figure 1b. Even if higher velocities (peak value: 8 mm/s) take place in the distributor region, particularly where the flow changes in direction, velocity values rapidly dampen as the fluid moves away from the distributor towards the sorbent. In the sorbent region, peak velocity spots are located immediately below the impact surface (range 0.354–0.356 mm/s). This value was found to be nearly insensitive either to permeability α and fluid viscosity μ, or to the axial position.

Mesoscale Study

The WSS patterns calculated at the fibers' surfaces for the three different cluster models are shown in figures 1c–e. Cluster A, owing to the perfect alignment of the different fiber layers, displayed quite a homogenous WSS pattern. Uneven WSS patterns were found for clusters B and C.

The calculated WSS frequency distribution was largely spread on low values: the main parameters of the distributions for clusters A, B and C (the median, the 90th percentile and the upper tail maximum) are summarized in table 1. The worse WSS distributions were found for clusters B and C, which were used as reference WSS cases for the microscale model.

Table 1. Relevant values of the WSS distributions obtained in the mesoscale study

Cluster ID	Wall elements number	Median WSS value (Pa)	90th percentile WSS value (Pa)	Maximum WSS value (Pa)
A	25,066	0.0657	0.364	0.824
B	21,423	0.140	1.04	4.91
C	16,756	0.154	0.935	2.89

Microscale Study

In the presence of a single PMB molecule grafted at the fiber wall, the fate of an endotoxin particle depends both upon its initial distance from the fiber wall and upon WSS. At each WSS (range: 0.06–5 Pa), there is a threshold value y_T, such that particles starting their motion at $y < y_T$ are captured by the PMB molecule; particles starting at $y > y_T$ are dragged away by the flow. In figure 2a the threshold distances for the S-LPS and ReLPS molecules are plotted versus the applied WSS values. The *y*-WSS plane is thus parted into two regions: a binding region (below the line) identifying the conditions for the endotoxin to be captured and bound to the PMB, and a nonbinding region identifying the conditions for the drag force to prevail.

In turn, in the presence of an entire surface layer of PMB molecules and in the absence of any other disturbing agent, the destination of an endotoxin particle is definitely to be attracted towards the PMB layer (fig. 2b). However, even a small increase of the particle's initial distance from the fiber wall implies a large magnification of the length that the particle has to travel next to the fiber's surface before being captured.

Discussion

Endotoxin adsorption in the Toraymyxin device relies on the capability of immobilized PMB to bind endotoxins with a high specificity and with a stable bond. The formation of this bond is governed by hydrophobic interactions in the short range (distances up to 1 nm) and by electrostatic interactions for farther distances, which yields extremely smaller forces. Conveying the endotoxin-rich plasma to come in contact with the fiber surface is therefore a crucial point for attaining an efficient endotoxin removal. In this work, we focused on blood and plasma hydrodynamic phenomena: starting from the analysis of global hydrodynamics, then zooming in down to the submillimetric and submicrometric scales. The transfer of data through scale levels was performed pessimistically (i.e. in favor of safety), so as to get a general view of the worst-case functionality

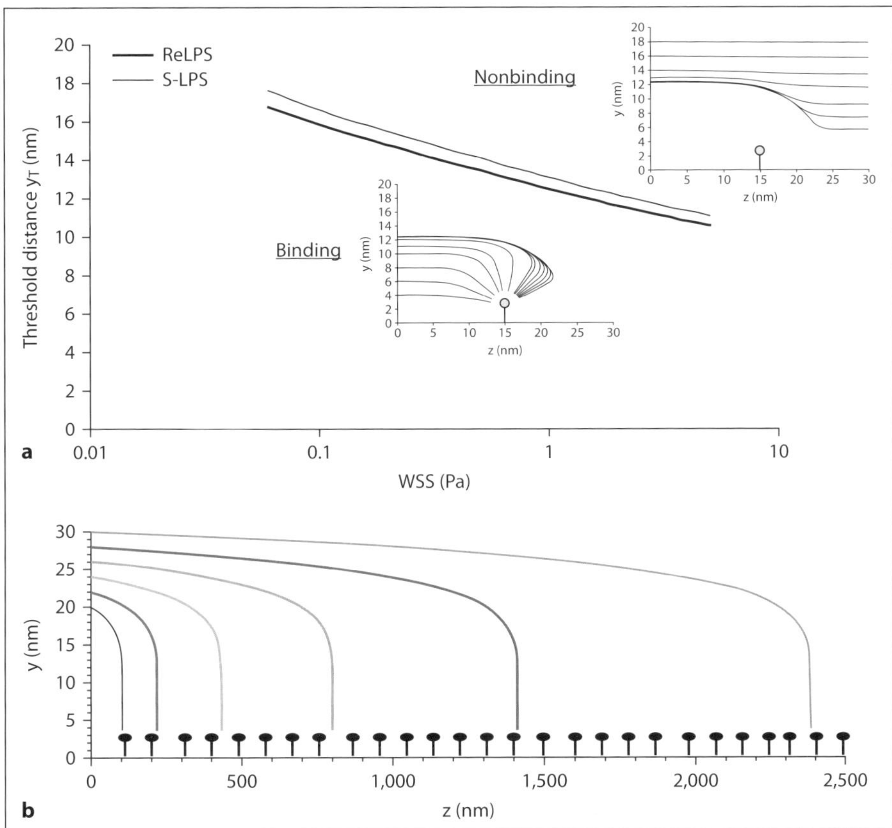

Fig. 2. Microscale study. **a** Relationship between the threshold distance and the applied WSS for the S-LPS and ReLPS molecules in the presence of a single PMB molecule at the fiber surface. Each line splits the plane in 2 regions: the binding region, and the nonbinding region. WSS values are plotted with a logarithmic scale. Insets depict the particle trajectories typical of the binding or nonbinding behavior. **b** The trajectories of endotoxin particles in the presence of an even distribution of PMB molecules at the fiber surface. Shown is the case of ReLPS particles starting their motion at a distance in the range 20–30 nm from the surface (step 2 nm) with WSS equal to 5 Pa.

of the device. With these presuppositions, according to our results, the grafted PMB is able to capture endotoxin molecules from a plasma layer extending tens of nanometers from the fiber surface. Although it is a small fraction of the thickness of the blood plasma sublayer, it is still far beyond the short-range interval that is characteristic of the ultimate stable intermolecular bond.

In our multiscale modeling framework, the primary goal of the macroscale study was to yield the peak velocity values in the sorbent region, which served as a worst-case input for the subsequent mesoscale model. For most of the sorbent volume, however, fluid velocities were far below the peak values, which were registered only immediately below the impact surface. The uniformity of the velocity field shows that the available sorbent volume is well exploited by the blood flow, confirming previous experimental observations [22]. Next, the study at the mesoscale level involved building realistic 3D CFD models of the flow around the fibers that constitute the sorbent material, so as to obtain a pessimistic range for the shear stresses in the fluid adjacent to the fiber walls, to be passed to the lower-scale analysis. The subsequent microscale study was performed choosing an observation scale at which the fluid could still be considered as a continuum, whereas endotoxin molecules could be modeled as suspended particles. This allowed us to study the local endotoxin adsorption phenomenon as a competition of the attraction exerted by the fiber-grafted PMB molecules versus the drag exerted by flow, the latter driving molecules to move together with the fluid stream (and hence, ultimately, out of the device). Our analysis was quite simplified with respect to reality. We disregarded the steric complexity of molecules, except for gross molecule dimensions (the equivalent radii of endotoxin molecules, the length of surface-grafted PMB molecules). Drag forces were expressed through the classical Stokes' law, despite the closeness of its limit of validity (the considered endotoxin molecules are 1 order of magnitude larger than water molecules). However, it is also thanks to simplicity that the developed model has the power of catching the essential characters of the phenomenon.

The potential for seizing endotoxins by the surface-grafted PMB molecules was investigated by tracking the trajectories of endotoxin molecules traveling with the flow at different distances above the surface. It is surprising to observe that with WSS values spanning 2 orders of magnitude (0.06–5 Pa), a single PMB molecule may capture endotoxin molecules traveling as far as 10–20 nm from the surface, which is a distant range for the molecular interaction. This, per se, is an indication of the way how hydrodynamics affect the adsorption potential. It has to be considered that at these distances the molecular attraction force is 8–10 orders of magnitude lower than the maximum attraction force. When an entire surface field of PMB molecules work synergistically, the seizing effect may even extend beyond the 20-nm distance, although for this to occur, the endotoxin molecule must travel as long as micrometers next to the fiber surface, which increases the probability that a disturbing effect (not considered here) will interfere with the molecular attraction.

Concluding Remarks

The whole multiscale framework depicted allows one to take a possible picture of the overall mechanisms influencing the antiendotoxinic effect of the Toraymyxin device. Close to the surface of fibers, even at relatively high WSS levels, an endotoxin-free layer of fluid is likely to form almost immediately. Above a certain distance (tens of nanometers), the endotoxin concentration would not be affected significantly. As a consequence of this, a concentration gradient would arise, promoting the diffusion of endotoxins towards the fiber wall. Hence, as far as the plasma sublayer of blood is considered, diffusion must be the major limiting factor for adsorption. This (together with the problem of sorbent saturation, which was not considered in this work), forces one to compensate with a huge active fiber area, which in the Toraymyxin device exceeds 500 m^2 by design. However, zooming out to a higher scale, convective effects would start playing a role in the transport of endotoxins from the bulk of the fluid towards the fiber surface. Therefore, the presence of a highly sheared local blood flow is appropriate because of the microconvective effects that may be generated by rotating erythrocytes or erythrocyte clusters in shear flow [23]. This may be obtained by enhancing the tortuosity of the local blood path through the fibers, even if it is recommended that the flow through the sorbent volume, taken as a whole, be free of any fluid dynamic irregularity, such as stagnation regions or preferential paths.

We investigated the mechanisms that underlie extracorporeal endotoxin removal by immobilized PMB through a multiscale computational study, ranging from the nanoscale, where the endotoxin/PMB molecular interactions take place, to the macroscale, i.e. the scale at which the treatment of the whole blood stream withdrawn from the patient is involved. Even if it is unavoidable to apply modeling simplifications at each observation scale, a multiscale approach has the power to provide a comprehensive picture, shedding light upon both the physics involved at each scale level and the mutual interactions of phenomena taking place at different levels.

Acknowledgements

The authors are grateful to Dr. Gualtiero Guadagni (Estor SpA, Milan, Italy) and Dr. Hisata Shoji (Toray Medical Inc., Tokyo, Japan) for supplying materials, technical drawings and design details.

References

1 Martin GS, Mannino DM, Eaton S, Moss M: The epidemiology of sepsis in the United States from 1979 through 2000. N Engl J Med 2003;348:1546–1554.

2 Murphy SL: Deaths: final data for 1998. Natl Vital Stat Rep 2000;48:1–105.

3 Friedman G, Silva E, Vincent JL: Has the mortality of septic shock changed with time? Crit Care Med 1998;26:2078–2086.

4 Balk RA: Pathogenesis and management of multiple organ dysfunction or failure in severe sepsis and septic shock. Crit Care Clin 2000;16:337–352, vii.

5 Opal SM, Gluck T: Endotoxin as a drug target. Crit Care Med 2003;31:S57–S64.

6 Palmer JD, Rifkind D: Neutralization of the hemodynamic effects of endotoxin by polymyxin B. Surg Gynecol Obstet 1974;138: 755–759.

7 From AH, Fong JS, Good RA: Polymyxin B sulfate modification of bacterial endotoxin: effects on the development of endotoxin shock in dogs. Infect Immun 1979;23:660–664.
8 Baldwin G, Alpert G, Caputo GL, Baskin M, Parsonnet J, Gillis ZA, Thompson C, Siber GR, Fleisher GR: Effect of polymyxin B on experimental shock from meningococcal and *Escherichia coli* endotoxins. J Infect Dis 1991;164:542–549.
9 Cooperstock MS: Inactivation of endotoxin by polymyxin B. Antimicrob Agents Chemother 1974;6:422–425.
10 Danner RL, Joiner KA, Rubin M, Patterson WH, Johnson N, Ayers KM, Parrillo JE: Purification, toxicity, and antiendotoxin activity of polymyxin B nonapeptide. Antimicrob Agents Chemother 1989;33: 1428–1434.
11 Wheeler AP: Bacterial peritonitis: innovative experimental treatment. Crit Care Med 1999;27:1055–1056.
12 Suzuki H, Nemoto H, Nakamoto H, Okada H, Sugahara S, Kanno Y, Moriwaki K: Continuous hemodiafiltration with polymyxin-B immobilized fiber is effective in patients with sepsis syndrome and acute renal failure. Ther Apher 2002;6:234–240.
13 Shoji H: Extracorporeal endotoxin removal for the treatement of sepsis: endotoxin adsorbtion cartridge (Toraymyxin). Ther Apher Dial 2003;7:108–114.
14 Kushi H, Miki T, Okamaoto K, Nakahara J, Saito T, Tanjoh K: Early hemoperfusion with an immobilized polymyxin B fiber column eliminates humoral mediators and improves pulmonary oxygenation. Crit Care 2005;9: R653–R661.
15 Cruz DN, Antonelli M, Fumagalli R, et al: Early use of polymyxin B hemoperfusion in abdominal septic shock. The EUPHAS randomized control trial. JAMA 2009;301: 2445–2452.
16 Nakamura T, Matsuda T, Suzuki Y, Shoji H, Koide H: Polymyxin B-immobilized fiber hemoperfusion in patients with sepsis. Dial Transplant 2003;32:602–607.
17 Nemoto H, Nakamoto H, Okada H, Sugahara S, Moriwaki K, Arai M, Kanno Y, Suzuki H: Newly developed immobilized polymyxin B fibers improve the survival of patients with sepsis. Blood Purif 2001;19:361–369.
18 Vincent JL, Laterre PF, Cohen J, Burchardi H, Bruining H, Lerma FA, Wittebole X, De Backer D, Brett S, Marzo D, Nakamura H, John S: A pilot-controlled study of a polymyxin B-immobilized hemoperfusion cartridge in patients with severe sepsis secondary to intra-abdominal infection. Shock 2005;23:400–405.
19 Fiore B, Soncini M, Vesentini S, Penati A, Visconti G, Redaelli A: Multiscale analysis of the Toraymyxin adsorption cartridge. Part II: computational fluid-dynamic study. Int J Artif Organs 2006;29:251–260.
20 Wells R: Syndromes of hyperviscosity. N Engl J Med 1970;283:183–186.
21 Vesentini S, Soncini M, Zaupa A, Silvestri V, Fiore GB, Redaelli A: Multi-scale analysis of the Toraymyxin adsorption cartridge. Part I: molecular interaction of polymyxin B with endotoxins. Int J Artif Organs 2006;29: 239–250.
22 Ronco C, Brendolan A, Scabardi M, Ronco F, Nakamura H: Blood flow distribution in a polymyxin coated fibrous bed for endotoxin removal. Effect of a new blood path design. Int J Artif Organs 2001;24:167–172.
23 Wang NL, Keller KH: Augmented transport of extracellular solutes in concentrated erythrocyte suspensions in Couette flow. J Colloid Interface Sci 1985;103:210–225.

Ing. Gianfranco B. Fiore
Department of Bioengineering, Politecnico di Milano
Piazza Leonardo da Vinci, 32
IT–20133 Milano (Italy)
Tel. +39 02 23993337, Fax +39 02 23993360, E-Mail gianfranco.fiore@polimi.it

Ronco C, Piccinni P, Rosner MH (eds): Endotoxemia and Endotoxin Shock: Disease, Diagnosis and Therapy. Contrib Nephrol. Basel, Karger, 2010, vol 167, pp 65–76

Endotoxin Removal by Polymyxin B Immobilized Cartridge Inactivates Circulating Proapoptotic Factors

Erica L. Martin · V. Marco Ranieri

Department of Anesthesiology and Critical Care, University of Turin, Ospedale S. Giovanni Battista-Molinette, Turin, Italy

Abstract

Background/Aims: Severe sepsis and septic shock continue to be major clinical challenges due to high associated mortality. Lipopolysaccharide (LPS) is a component of the cell membrane of Gram-negative bacteria, and is believed to initiate septic-induced signaling, inflammation and organ damage, including acute renal failure. Polymyxin B (PMX-B) hemoperfusion of septic patients can improve survival and decreasing organ dysfunction by removing circulating LPS. Unfortunately, some clinicians have been slow to adopt this novel therapy due to the lack of understanding of the cellular mechanisms involved in this treatment. Apoptosis, or programmed cell death, is known to contribute to acute renal failure and overall organ dysfunction during sepsis, and can be activated by LPS-initiated signaling pathways. Therefore, the protective renal effects associated with PMX-B hemoperfusion of septic patients may result from alterations in cellular apoptosis. This chapter will review recent data regarding the role of apoptosis prevention in the mechanism leading to the improved outcome and decreased acute renal failure associated with PMX-B hemoperfusion during sepsis. **Methods:** Blood was collected, upon inclusion and following 72 h, from conventionally treated patients and patients receiving two PMX-B hemoperfusion treatments. Plasma was subsequently used to stimulate renal tubule cells or glomerular podocytes to assess their ability to induce apoptosis. **Results:** All plasma collected upon inclusions, as well as plasma from conventionally treated patients at 72 h, significantly increased apoptosis, while plasma collected from patients following PMX-B treatment induced significantly less apoptosis than time 0 or conventionally treated controls. This decreased proapoptotic signal resulted from decreased extrinsic and intrinsic apoptotic signaling determined by decreased caspase activity, Fas expression and Bax/Bcl-2 balance. **Conclusion:** The protective effects of extracorporeal therapy with PMX-B on the development of acute renal failure result, in part, through its ability to reduce the systemic proapoptotic activity of septic patients on renal cells.

Severe sepsis and septic shock, involving wide-spread systemic inflammation and infection, are serious medical conditions that continue to plague intensive care units [1]. Furthermore, despite intensive research and progress in our understanding of the mechanisms involved in the pathology of sepsis, its associated mortality remains at 40–60%, often resulting from the progression of sepsis to multiple organ failure [1]. One of the organs prone to septic-induced failure is the kidney, cumulating in acute renal failure (ARF) in 23% of patients with severe sepsis and 51% of patients with septic shock [2]. In addition, the presence of ARF significantly increases the probability of mortality to 70% [2].

Lipopolysaccharide (LPS), a component of the Gram-negative bacterial cell wall, is believed to be one of the driving mechanic forces behind the induction of ARF during sepsis [3]. LPS induces many of its effects through the activation of the Toll-like receptor 4 signaling cascade, involving mitogen-activated protein kinases and NFκB, which in turn augment the inflammatory response and induce cellular damage and organ dysfunction [4]. Removal of circulating LPS by hemoperfusion through an immobilized polymyxin B (PMX-B) column has been shown to improve septic outcome and reduce the incidence of septic-induced ARF [5–7]; however, some physicians are hesitant to initiate the widespread application of this novel therapy without a better understanding of the cellular mechanisms leading to its beneficial effects.

Apoptosis

The process of apoptosis involves the active elimination of cells through the initiation of programmed cell death [8]. This process is normally triggered when cells are damaged beyond repair and can be elicited by the cell itself, its surrounding tissue or via the immune system [8]. The process of apoptosis is characterized by specific changes in cellular shape and organization, including cell blebbing, loss of membrane asymmetry and attachment, cell shrinkage, nuclear fragmentation, chromatin condensation, chromosomal DNA fragmentation, and the presence of apoptotic bodies [9].

Ultimately, this process ends in the disposal of all cellular debris through phagocytosis in order to prevent an inflammatory reaction or organ damage [9].

Apoptotic Pathways

Apoptosis occurs primarily through three different signaling pathways: the extrinsic (death receptor) pathway, the intrinsic (mitochondrial) pathway and the endoplasmic reticulum (stress-induced) pathway [8].

Death receptors of the extrinsic pathway belong to the tumor necrosis factor (TNF) superfamily of membrane receptors [8]. While several death receptors have been described, the Fas antigen and TNF receptor 1 (TNF-R1) are the most dominant and best described activators of this pathway [8, 10]. They are activated by Fas ligand (FasL) and TNF-α, respectively, which induces trimerization to create the death-induced signal complex, and recruit the adaptor proteins FADD (for Fas) or TRADD (for TNF-R1) [8, 10]. This complex can then bind the activated death domains of procaspase-8, resulting in its cleavage and activation [8]. Caspase-8, in turn, cleaves caspases-3, -6 and -7, which cleaves DNA leading to apoptosis [8]. Additionally, caspase-8 can activate the proapoptotic molecule BID, which induces cross-talk and apoptosis through the intrinsic pathways [11].

The intrinsic apoptotic pathway can be triggered by cellular or mitochondrial stress, including changes in growth factors, cytokines, steroids, reactive oxygen species, nitric oxide and heat shock proteins [8]. These signals induce the translocation of proapoptotic members of the Bcl-2 family (i.e. BID, Bax) from the cytosol to the mitochondrial membrane, where they reduce the mitochondrial membrane potential. Under normal conditions, antiapoptotic Bcl-2 family members (Bcl-2, Bcl-xL) block the release of cytochrome c from the mitochondria [8]. However, when the balance of Bcl-2 family members shifts from antiapoptotic to proapoptotic, this induces the release of large amounts of destructive cytochrome c from the mitochondria [8]. Upon its release, cytochrome c combines with adenosine triphosphate and the enzyme Apaf to active caspase-9 to form an apoptosome, which then cleaves and activates caspases-3, -6 and -7, which similarly to the extrinsic pathway, cleave DNA resulting in cell death [8].

The endoplasmic reticulum stress-induced pathway of apoptosis is the least understood and can be induced by many cellular stresses, such as hypoxia, oxidative injury, infection and calcium disturbances [12]. Each of these disturbances leads to the accumulation of unfolded proteins within the endoplasmic reticulum, causing the dissociation of the glucose-regulated protein (78 kDa) from its three endoplasmic reticulum stress receptors (PERK, ATF6, IRE1), thereby initiating the unfolded protein response [12]. This dissociation of the glucose-regulated protein (78 kDa) activates downstream signaling pathways that can either (1) induce apoptosis directly through the activation of caspase-12, or its possible human homologue caspase-4; (2) induce activation of the intrinsic mitochondrial pathway; or (3) initiate caspase-independent cell death [12].

Apoptotic Effects on Tissue/Organ Function

The presence of significant apoptosis can effect the activity and overall function of the affected tissue or organ, either by decreasing the critical number of viable cells or inhibiting communication between cells [9, 13]. During sepsis, apoptosis occurs in different cell types, each inducing various forms of dysfunction [9]. For example, immune suppression has been shown to result from apoptosis in lymphocytes, the thymus and/or spleen; vascular permeability can be caused by endothelial apoptosis; cardiac depression has been shown to be caspase-dependent; and acute lung injury is correlated to the amount of Fas/FasL-induced apoptosis in alveolar lung cells [14].

Septic-induced ARF has also been specifically associated with the induction of apoptotic mechanisms in various renal cell types, including tubule cells, glomerular endothelial cells and podocytes [15–18]. Renal tubule cells exposed to plasma from septic patients display increased apoptosis and disturbed permeability, inflammation, adhesion and tubular formation [17, 19]. In addition, this proapoptotic signal correlates with the severity of proteinuria in patients [17]. Similarly, renal endothelial cells exposed to plasma from patients with ARF show increased Fas expression and apoptosis, which can be blocked by either overexpression of the antiapoptotic mediator Bcl-2, or by inhibition of caspases [20]. Furthermore, a separate study found that both LPS and TNF-α induce apoptosis in glomerular endothelial cells, which are essential for the regulation of glomerular ultrafiltration, and that this response is caspase-dependent [18]. Podocytes, or visceral epithelial cells, are involved in glomerular filtration and are also stimulated to undergo apoptosis when exposed to plasma from septic patients [19].

LPS Activation of Apoptosis

The mechanisms through which LPS induces apoptosis is not entirely clear; however, several lines of evidence indicate that the extrinsic and intrinsic pathways play a dominant role in this process. Several studies have convincingly shown that sepsis-induced renal apoptosis is dependent on the activity of caspases, either through the use of caspase inhibitors or caspase-deficient mice [16, 21, 22]. Additionally, Guo et al. [21] demonstrated that caspase inhibition decreased renal inflammation, despite the fact that the process of apoptosis is believed to be noninflammatory, suggesting that the blockade of caspase-dependent apoptosis may also decrease nonapoptotic renal injury. The extrinsic pathway is believed to initiate the process of apoptosis by LPS since both LPS and TNF, produced by LPS signaling, upregulate the mRNA of Fas and FasL in renal tubule cells [23]. Therefore, LPS activates both the Fas and TNF-R1 death receptors, thereby triggering the extrinsic pathway of apoptosis. In addition, the inflammatory response induced by LPS produces a vast array of cytokines capable of stimulating the intrinsic pathway through altering the balance of the Bcl-2 family [19].

PMX-B Hemoperfusion and Apoptosis

Since apoptosis has been shown to play an important role in the development of septic-induced ARF, and LPS, a key mediator in the pathogenesis of sepsis, is known to initiate apoptosis in various renal cells, it was hypothesized that PMX-B hemoperfusion therapy, which removes circulating LPS in septic patients, may reduce ARF through the reduction of proapoptotic signaling [19]. In order to test this hypothesis, plasma from septic patients who received either conventional treatment alone or conventional treatment with two PMX-B hemoperfusion sessions were tested on renal tubular cells and glomerular podocytes.

Patients were eligible to participate if they had a confirmed or suspected Gram-negative infection, at least three indications of systemic inflammatory response syndrome, at least one organ dysfunction, and could be randomized within 24 h of matching study criteria [19]. Once enrolled with informed consent, plasma samples were taken at time 0. Patients randomized to PMX-B treatment then received one hemoperfusion session upon inclusion and a second treatment at 24 h. For both patient groups, plasma was also collected at 2, 26 and 72 h following study inclusion. Additionally throughout the 72-hour study period, demographic and clinical data were recorded.

At enrollment, the patient groups were not significantly different in any way; however, in agreement with other PMX-B hemoperfusion studies, patients in the PMX-B treatment group showed a significant decrease in their SOFA scores over 72 h, while conventionally treated patients showed no improvement in SOFA scores [19]. Furthermore, a similar pattern was observed for RIFLE scores, with the PMX group displaying significant improvement, which was not observed in the conventionally treated group. In addition, the need for renal replacement therapy was significantly greater in the conventionally treated patients versus PMX-B hemoperfused patients [19]. Together, these data indicate that the patient plasma collected in this study reflects other historical studies, showing protection against the development of septic-induced ARF following PMX-B hemoperfusion.

Renal Cytotoxicity

Cellular viability and cytotoxicity was measured using an XTT-based assay, showing that, in both glomerular podocytes and tubular cells, plasma from PMX-B-treated patients induces significantly less cytotoxicity compared to pretreatment or conventionally treated control patients [19]. This effect was observed immediately following both PMX-B hemoperfusion sessions and 72 h following enrollment, indicating that the protection following PMX-B therapy remains constant over time. While this measurement gives a general assessment of cellular cytotoxicity, it does not specify the type of cellular death.

TUNEL Apoptotic Assessment

In order to determine the specific role of apoptosis, renal tubular cells exposed to PMX-B or conventionally treated plasma were assessed using a terminal

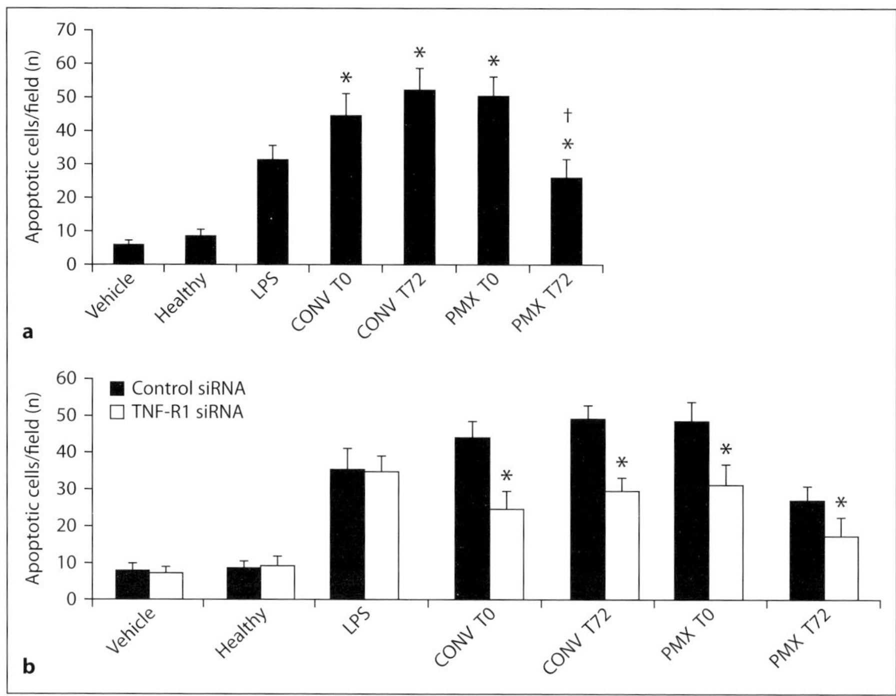

Fig. 1. **a** Tubular apoptosis (TUNEL) induced by conventional (CONV) or PMX plasmas. All PMX and CONV plasmas induced a significant increase of tubular apoptosis (* $p < 0.05$). Incubation with PMX T72 plasma resulted in a significant decrease of apoptosis compared to PMX T0 plasma († $p < 0.05$). **b** Evaluation of tubular apoptosis (TUNEL) using short interfering RNA (siRNA) for TNF-R1. Compared to control siRNA, a significant decrease of tubular apoptosis was observed in siRNA TNFR1 tubular cells incubated with CONV and PMX plasma (* $p < 0.05$).

deoxynucleotidyl transferase dUTP nick end labeling (TUNEL) assay (fig. 1). Plasma from both groups of patients at time 0 induced significant tubular cell apoptosis compared to healthy controls [19]. Interestingly, while plasma from 72 h of conventionally treated patients continued to induce a similar level of apoptosis to that of time 0, plasma taken at 72 h from PMX-B-treated septic patients produced significantly less apoptosis than time 0 or conventionally treated patients, thereby indicating that reduced apoptosis contributes to the cellular protection observed following PMX-B hemoperfusion [19]. Furthermore, when PMX-B was added to the plasma culture at a dose established to inhibit LPS biological activity without causing tubular cell death, there was an apoptotic effect for both patient groups at time 0, and of conventionally treated patients at time 72, while PMX-B addition had no effect on the PMX-B hemoperfused group at 72 h [19]. This indicates that the protective effect observed in the plasma from PMX-B patients at 72 h results from decreased LPS biological activity.

Caspase Activity

Most pathways of apoptosis result in the activation of caspases, which cut DNA and ultimately induce cell death. As expected, plasma from time 0 of both conventional and PMX-B groups induced increase caspase-3, -8 and -9 activity, indicating that the renal apoptosis caused by septic plasma occurs through caspase-dependent pathways [19]. Furthermore, in agreement with the TUNEL results, plasma taken at 72 h from conventionally treated patients induced similar levels of caspase activity to time 0, whereas plasma from PMX-B treated patients induced significantly less activity of all three caspases as compared to time 0 or conventionally treated controls [19]. Although caspase-3 is common to both the extrinsic and intrinsic pathways, the fact that PMX-B treatment reduced the activity of both caspase-8, which is activated by the death receptors, and caspase-9, which is activated by mitochondrial cytochrome c, implies that both the extrinsic and intrinsic pathways are involved in this mechanism.

Extrinsic Pathway

Since caspase-8 activity was altered by PMX-B hemoperfusion therapy of septic patients, it is likely that this treatment protects against apoptosis through an alteration of the extrinsic pathway. As described above in this chapter, caspase-8 is cleaved and activated by the apoptotic death receptors TNF-R1 and Fas [8]. Our work has additionally shown that while both TNF-R1 and Fas signaling are likely involved, the protection observed through PMX-B hemoperfusion is likely dominated by the Fas pathway. This conclusion is based on the observation that while blockade of TNF-R1 signaling, through the use of short interfering RNA (siRNA) targeting this receptor, significantly reduced the proapoptotic effect of septic plasma taken at time 0, it also significantly reduced the apoptosis observed in response to plasma from PMX-B treated patients [19]. These results indicate that while PMX-B therapy reduces TNF-R1-induced apoptosis, it does not completely block this signaling pathway. In contrast, the high levels of Fas expression of tubular cells following exposure to septic plasma was drastically reduced when plasma was taken from PMX-B-treated patients [19], thereby suggesting a larger role for the involvement of Fas signaling in the mechanisms leading to the protection resulting from PMX-B therapy.

Intrinsic Pathway

Although LPS is more clearly linked with the activation of the extrinsic apoptotic pathway, the observed alteration of caspase-9 activity in response to PMX-B hemoperfusion of septic patients suggests an additional role of the intrinsic pathway [8]. Since the activation of apoptosis via the intrinsic pathway is dependent on the balance between the proapoptotic and antiapoptotic members of the Bcl family, our study reports the ratio of Bax, a proapoptotic protein, to Bcl-2, an antiapoptotic protein [19]. This analysis shows that while plasma from conventionally treated septic patients induced a higher Bax/Bcl-2 ratio over the 72-hour study period, plasma from patients who received PMX-B hemoperfusion had a significantly lower Bax/Bcl-2 ratio [19], indicating a role for intrinsic apoptotic signaling in the mechanisms causing PMX-B-induced improved cellular viability.

Additional Nonapoptotic Mechanisms

In addition to altering apoptosis, improved outcome resulting from PMX-B hemoperfusion during sepsis may result from other cellular mechanisms, including impaired cellular orientation, improved reabsorption capacity, and reduced inflammation and permeability [19, 24–26].

To obtain proper tubular formation, renal cells must obtain proper apical-basal orientation. When renal tubule cells were stimulated with septic plasma, cells lost this orientation and, thus, were unable to obtain tubular formation; however, when stimulated with plasma from septic PMX-B treated patients, renal cells were able to adhere to matrix and restore tubular structures [19].

An important function of tubule renal cells is the reabsorption of glomerular-filtered low molecular weight proteins and molecules, preventing their loss

Fig. 2. a ELISA evaluation of caspase-3, -8 and -9 activity on tubular cells cultured with conventional (CONV) or PMX plasma. PMX and CONV plasmas induced increased activity of all caspases (*$p < 0.05$ vs. healthy plasma). A significant decrease of all caspase activities was found with PMX T72 plasma compared to PMX T0 († $p < 0.05$ PMX T72 vs. PMX T0); however, caspase-3 and -9 activities remained significantly higher than healthy plasma (* $p < 0.05$ PMX T72 vs. healthy plasma). **b** Representative images of FACS and immunofluorescence (insets) analysis of Fas (CD95) expression on tubular cells. PMX T0 and CONV T0 and T72 plasmas all induced a marked up-regulation of Fas, which was significantly reduced in the presence of PMX T72 plasma. Magnification: 400×. **c** Representative Western blot analysis of the mitochondrial proteins Bax and Bcl-2 in tubular cells, and related densitometric analysis expressed as Bax/Bcl-2 ratio. PMX T0, CONV T0 and CONV T72 plasmas induced a marked upregulation of the Bax/Bcl2 ratio that was reduced in the presence of PMX T72 plasma. (lane 1: vehicle; lane 2: healthy; lane 3: PMX T0; lane 4: PMX T72; lane 5: CONV T0; lane 6: CONV T72).

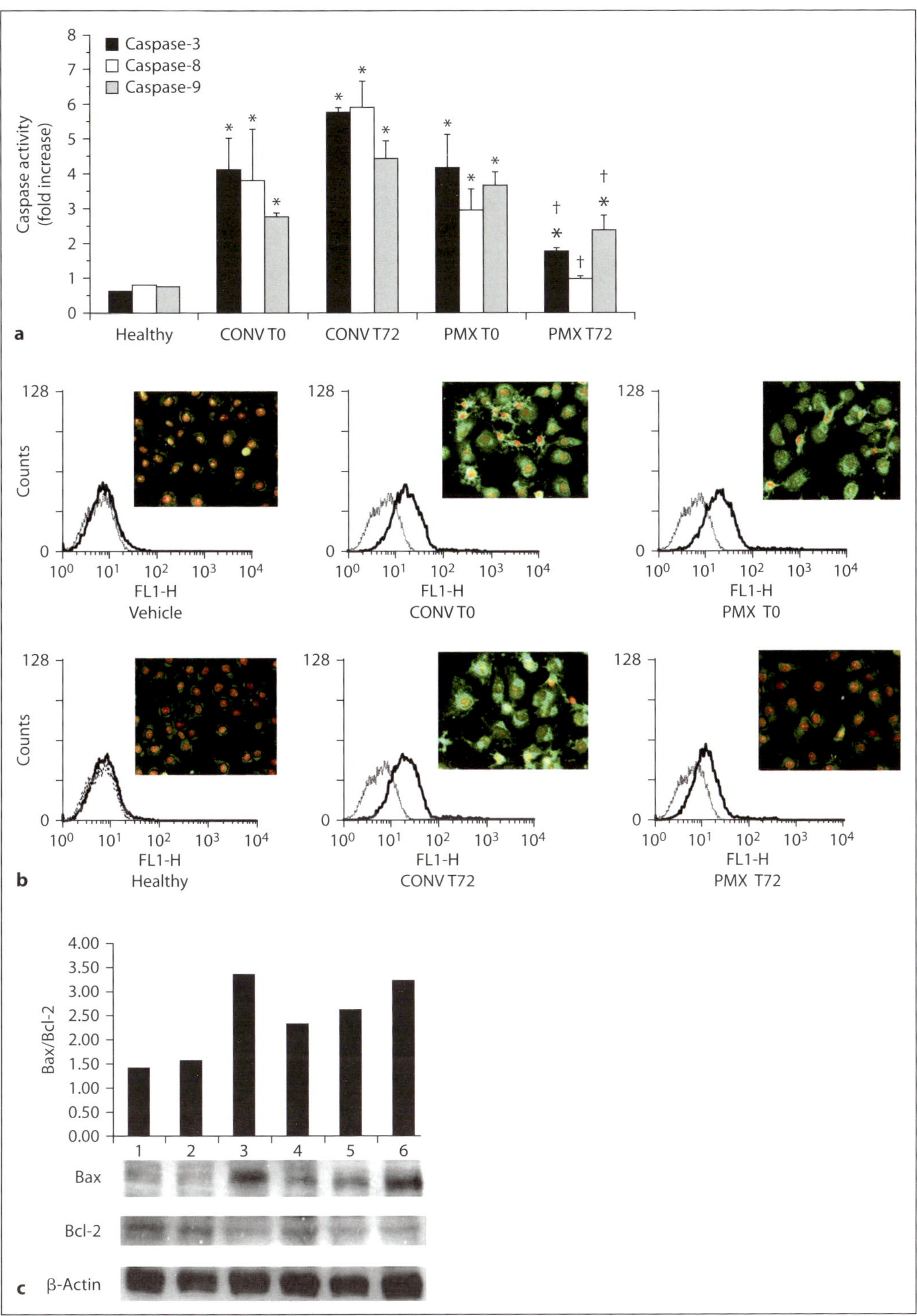

Caspase-3
Caspase-8
Caspase-9
Caspase activity (fold increase)
Healthy
CONV T0
CONV T72
PMX T0
PMX T72
a
Counts
FL1-H
Vehicle
CONV T0
PMX T0
Healthy
CONV T72
PMX T72
b
Bax/Bcl-2
Bax
Bcl-2
β-Actin
c

in the urine. Megalin is one of the transport receptors found on the luminal surface of renal proximal tubules [25]. Our study found that exposure of renal tubule cells to septic plasma significantly reduced megalin expression, while exposure to plasma from PMX-B-treated patients displayed megalin levels similar to healthy or vehicle-treated controls [19], suggesting a complete blockade of this reabsorption impairment.

Generally, renal cellular injury is often characterized by the presence of inflammation and increased cellular permeability [24, 26], which through the altered expression of CD40, ICAM-1, B7–1 and nephrin, as well as differences in resistance and albumin permeability, was observed in both renal tubule cells and glomerular podocytes upon exposure to plasma from septic patients [19]. In contrast, renal cells exposed to PMX-B-treated plasma induced significantly less inflammation and permeability [19], suggesting that these mechanisms may contribute to the improved outcome associated with PMX-B hemoperfusion.

Supportive Evidence in Animals

In support of these findings, a recent animal study performed using the cecal ligation and perforation model in Sprague-Dawley rats has confirmed the conclusion that PMX-B hemoperfusion can reduce the proapoptotic effect on kidneys [27]. In this study, rats underwent cecal ligation and perforation to induce sepsis, then after 24 h received a 1-hour PMX-B hemoperfusion treatment [27]. At 1 h after the external circulation, animals were sacrificed and assessed for apoptosis in the lung, liver and kidneys by TUNEL assay [27]. The results showed that rats who received the PMX-B treatment had significantly less apoptosis in renal tubule cells compared to nontreated controls [27]. While their results did not show any significant alteration in lung or liver apoptosis due to the PMX-B hemoperfusion [27], this may result from the timing at which apoptosis was analyzed or the fact that PMX-B therapy was not given until 24 h after cecal ligation and perforation. Therefore, while this study clearly demonstrates a role of renal apoptosis, it cannot exclude the possibility of the contribution of apoptosis in other organs.

Overall Conclusions

In summary, apoptosis plays an important role in multiple organ failure and, in particular, ARF during systemic sepsis. Furthermore, circulating LPS during sepsis can initiate apoptotic programmed cell death, thereby suggesting that removal of LPS through PMX-B hemoperfusion may improve renal function and overall patient outcome through decreasing the systemic proapoptotic signal. Recent experimental studies and results from human renal tubular cells and glomerular podocytes confirm the role of apoptosis as part of the cellular mechanisms leading to a decrease in ARF following PMX-B treatment. Through further understanding of these mechanisms involved in PMX-B hemoperfusion protection, we aim to improve its clinical acceptance and thereby increase its application to patients at risk.

References

1 Dremsizov TT, Kellum JA, Angus DC: Incidence and definition of sepsis and associated organ dysfunction. Int J Artif Organs 2004;27:352–359.

2 Schrier RW, Wang W: Acute renal failure and sepsis. N Engl J Med 2004;351:159–169.

3 Cunningham PN, Wang Y, Guo R, He G, Quigg RJ: Role of Toll-like receptor 4 in endotoxininduced acute renal failure. J Immunol 2004;172:2629–2635.

4 Salomao R, Martins PS, Brunialti MK, Fernandes ML, Martos LS, Mendes ME, Gomes NE, Rigato O: TLR signaling pathway in patients with sepsis. Shock 2008;30(Suppl 1):73–77.

5 Cruz DN, Antonelli M, Fumagalli R, Foltran F, Brienza N, Donati A, Malcangi V, Petrini F, Volta G, Bobbio Pallavicini FM, Rottoli F, Giunta F, Ronco C: Early use of polymyxin B hemoperfusion in abdominal septic shock: the EUPHAS randomized controlled trial. JAMA 2009;301:2445–2452.

6 Cruz DN, Perazella MA, Bellomo R, de Cal M, Polanco N, Corradi V, Lentini P, Nalesso F, Ueno T, Ranieri VM, Ronco C: Effectiveness of polymyxin B-immobilized fiber column in sepsis: a systematic review. Crit Care 2007;11:R47.

7 Nakamura T, Kawagoe Y, Matsuda T, Ueda Y, Koide H: Effects of polymyxin B immobilized fiber on urinary N-acetyl-beta-glucosaminidase in patients with severe sepsis. ASAIO J 2004;50:563–567.

8 Wesche-Soldato DE, Swan RZ, Chung CS, Ayala A: The apoptotic pathway as a therapeutic target in sepsis. Curr Drug Targets 2007;8:493–500.

9 Hotchkiss RS, Swanson PE, Freeman BD, Tinsley KW, Cobb JP, Matuschak GM, Buchman TG, Karl IE: Apoptotic cell death in patients with sepsis, shock, and multiple organ dysfunction. Crit Care Med 1999;27: 1230–1251.

10 Sheikh MS, Huang Y: Death receptor activation complexes: it takes two to activate TNF receptor 1. Cell Cycle 2003;2:550–552.

11 Strasser A, Jost PJ, Nagata S: The many roles of FAS receptor signaling in the immune system. Immunity 2009;30:180–192.

12 Kim I, Xu W, Reed JC: Cell death and endoplasmic reticulum stress: disease relevance and therapeutic opportunities. Nat Rev Drug Discov 2008;7:1013–1030.

13 Hotchkiss RS, Nicholson DW: Apoptosis and caspases regulate death and inflammation in sepsis. Nat Rev Immunol 2006;6:813–822.

14 Papathanassoglou ED, Moynihan JA, Ackerman MH: Does programmed cell death (apoptosis) play a role in the development of multiple organ dysfunction in critically ill patients? a review and a theoretical framework. Crit Care Med 2000;28:537–549.

15 Taguchi T, Uchida H, Kiyokawa N, Mori T, Sato N, Horie H, Takeda T, Fujimoto J: Verotoxins induce apoptosis in human renal tubular epithelium derived cells. Kidney Int 1998;53:1681–1688.

16 Ueda N, Kaushal GP, Shah SV: Apoptotic mechanisms in acute renal failure. Am J Med 2000;108:403–415.

17 Mariano F, Cantaluppi V, Stella M, Romanazzi GM, Assenzio B, Cairo M, Biancone L, Triolo G, Ranieri VM, Camussi G: Circulating plasma factors induce tubular and glomerular alterations in septic burns patients. Crit Care 2008;12:R42.

18 Messmer UK, Briner VA, Pfeilschifter J: Tumor necrosis factor-alpha and lipopolysaccharide induce apoptotic cell death in bovine glomerular endothelial cells. Kidney Int 1999;55:2322–2337.

19 Cantaluppi V, Assenzio B, Pasero D, Romanazzi GM, Pacitti A, Lanfranco G, Puntorieri V, Martin EL, Mascia L, Monti G, Casella G, Segoloni GP, Camussi G, Ranieri VM: Polymyxin-B hemoperfusion inactivates circulating proapoptotic factors. Intensive Care Med 2008;34:1638–1645.
20 Mitra D, Kim J, MacLow C, Karsan A, Laurence J: Role of caspases 1 and 3 and Bcl-2-related molecules in endothelial cell apoptosis associated with thrombotic microangiopathies. Am J Hematol 1998;59: 279–287.
21 Guo R, Wang Y, Minto AW, Quigg RJ, Cunningham PN: Acute renal failure in endotoxemia is dependent on caspase activation. J Am Soc Nephrol 2004;15:3093–3102.
22 Wang W, Faubel S, Ljubanovic D, Mitra A, Falk SA, Kim J, Tao Y, Soloviev A, Reznikov LL, Dinarello CA, Schrier RW, Edelstein CL: Endotoxemic acute renal failure is attenuated in caspase-1-deficient mice. Am J Physiol Renal Physiol 2005;288:F997–F1004.
23 Ortiz-Arduan A, Danoff TM, Kalluri R, Gonzalez-Cuadrado S, Karp SL, Elkon K, Egido J, Neilson EG: Regulation of Fas and Fas ligand expression in cultured murine renal cells and in the kidney during endotoxemia. Am J Physiol 1996;271:F1193–F1201.
24 Reiser J, von Gersdorff G, Loos M, Oh J, Asanuma K, Giardino L, Rastaldi MP, Calvaresi N, Watanabe H, Schwarz K, Faul C, Kretzler M, Davidson A, Sugimoto H, Kalluri R, Sharpe AH, Kreidberg JA, Mundel P: Induction of B7-1 in podocytes is associated with nephrotic syndrome. J Clin Invest 2004;113:1390–1397.
25 Christensen EI, Birn H: Megalin and cubilin: synergistic endocytic receptors in renal proximal tubule. Am J Physiol Renal Physiol 2001;280:F562–F573.
26 De Gaudio AR, Adembri C, Grechi S, Novelli GP: Microalbuminuria as an early index of impairment of glomerular permeability in postoperative septic patients. Intensive Care Med 2000;26:1364–1368.
27 Ito M, Kase H, Shimoyama O, Takahashi T: Effects of polymyxin B-immobilized fiber using a rat cecal ligation and perforation model. ASAIO J 2009;55:246–250.

Erica L. Martin
Dipartimento di Anestesiologia e Rianimazione, Università di Torino, Ospedale S. Giovanni Battista-Molinette
Corso A.M. Dogliotti 14, IT–10126 Torino (Italy)
Tel. +39 011 633 4005, Fax +39 011 696 0448, E-Mail ericaleanne.martin@unito.it

Ronco C, Piccinni P, Rosner MH (eds): Endotoxemia and Endotoxin Shock: Disease, Diagnosis and Therapy. Contrib Nephrol. Basel, Karger, 2010, vol 167, pp 77–82

Polymyxin-B Hemoperfusion and Endotoxin Removal: Lessons from a Review of the Literature

Dinna N. Cruz[a,c] · Massimo de Cal[a,c] · Pasquale Piccinni[b] · Claudio Ronco[a,c]

Departments of [a]Nephrology, Dialysis and Transplantation and [b]Anesthesiology and Intensive Care Medicine, San Bortolo Hospital, and [c]International Renal Research Institute (IRRIV), Vicenza, Italy

Abstract

Sepsis involves a complex interaction between bacterial toxins and the host immune system. Endotoxin, a component of the outer membrane of Gram-negative bacteria, is involved in the pathogenesis of sepsis producing proinflammatory cytokines and activating the complement system, and is thus an ideal potential therapeutic target. Direct hemoperfusion using polymyxin B-immobilized fiber column (PMX-F) has been shown to bind and neutralize endotoxin in both in vitro and in vivo studies. Therefore, this extracorporeal therapy with PMX-F can potentially interrupt the biological cascade of sepsis. A systematic review of the published literature found positive effects of PMX-F on blood pressure and dopamine/dobutamine use, the PaO_2/FiO_2 ratio, endotoxin removal, and mortality. It should be noted, however, that many of the analyzed studies were of suboptimal quality, which may then exaggerate the magnitude of these effects. Since this meta-analysis, other studies have been published including a multicenter randomized controlled trial on abdominal septic shock. In this study, PMX-F, when added to conventional therapy, significantly improved hemodynamics and organ dysfunction, and reduced 28-day mortality in this targeted population. There is clear biological rationale for endotoxin removal in the clinical management of severe sepsis and septic shock. The current literature seems to provide some support for this premise, and provides the basis for further rigorous study.

Sepsis is characterized by an overwhelming systemic inflammatory response and subsequent immune dysfunction. It is a major cause of death in intensive care units, with an estimated incidence in the United States of 750,000 cases per year and a mortality rate of 25–80% [1]. Sepsis involves a complex interaction between bacterial toxins and the host immune system. Endotoxin, a component of the outer membrane of Gram-negative bacteria, is involved in the pathogenesis of sepsis producing proinflammatory cytokines, including TNF-α and IL-1, which play roles as mediators in inflammatory responses and organ injuries [2, 3]. Moreover, endotoxin activates complements and coagulation factors, and is an ideal potential therapeutic target to treat septic shock [2]. However, lack of demonstrable clinical benefit with antiendotoxin or anticytokine therapy has shifted interest to extracorporeal therapies to reduce circulating levels of septic mediators.

Polymyxin B (PMX) is a cationic cyclic polypeptide antibiotic which binds with high affinity to endotoxin, neutralizing its effects. However, it has significant nephrotoxic and neurotoxic effects, and these toxicities preclude its systemic use. This subsequently led to the development of an adsorptive cartridge in which PMX is covalently bound to polystyrene fibers (PMX-F) [4]. This design capitalizes on the endotoxin-neutralizing effects of PMX while minimizing systemic toxicity. Direct hemoperfusion through this PMX-F column has been shown to bind and neutralize endotoxin in both in vitro and in vivo studies [4]. Therefore, the extracorporeal therapy by PMX-F can reduce circulating endotoxin levels and potentially interrupt the biological cascade of sepsis. The PMX-F column has been in clinical use in Japan since 1994 for patients with endotoxemia or suspected Gram-negative infection who fulfill the conditions of systemic inflammatory response syndrome (SIRS) and have septic shock requiring vasoactive agents. Since 1994, more than 70,000 patients are said to have received this treatment [Toray Industries, unpubl. data].

A Systematic Review of the Literature

Several studies have demonstrated efficient removal of endotoxin with PMX-F therapy as well as suppression of TNF-α production. However, despite the well-documented capacity to lower blood endotoxin levels, the impact of this therapy on clinical endpoints remains unclear. In early studies, PMX-F therapy appears to effectively reduce endotoxin levels and have some positive effects on blood pressure, use of vasoactive agents, gas exchange and short-term mortality. However, many studies were small and underpowered, and therefore inconclusive. In 2007, we performed a systematic review on the effectiveness of PMX-F in sepsis to summarize the available clinical literature [5, 6]. This meta-analysis covered 9 randomized controlled trials and 19 observational studies which included more than 1,400 patients treated in 7 countries. The results of this analysis and some of the included studies are briefly summarized below.

In 9 randomized controlled trials and 10 observational studies, levels of circulating endotoxin decreased by 33–80% from baseline levels when patients were treated with PMX-F [6]. In addition to its capacity to lower endotoxin levels, reduced levels of other mediators such as IL-6 [7–9], IL-10 [8, 9], IL-18 [10], TNF-α [9, 11], metalloproteinase-9 [12], plasminogen activator inhibitor-1 [9, 11, 13, 14], neutrophil elastase [13, 14], platelet factor-4 [15], β-thromboglobulin [15], soluble P-selectin [15] and endogenous cannabinoids, such as anandamide [16], have also been reported after PMX-F therapy.

The effect of PMX-F therapy on pulmonary function was ascertained in a pooled analysis of 7 studies [6]. Overall, the PaO_2/FiO_2 ratio increased by 32 units (95% CI: 23–41 units) after PMX-F, suggesting an improvement in pulmonary function. This finding was also seen in later studies. PMX-F therapy improved the PaO_2/FiO_2 ratio in patients with acute lung injury or acute respiratory distress syndrome caused by sepsis [17], and this appeared to bc related to reduction in the blood neutrophil elastase and IL-8 levels [13].

In terms of hemodynamics, the mean arterial pressure (MAP) increased by 26% (range: 14–42%) after PMX-F [6]. The mean weighted difference was 19 mm Hg (95% CI: 15–22 mm Hg). The degree to which the MAP improved depended on the severity of hypotension of the patients. Studies which enrolled patients with lower baseline blood pressure demonstrated a bigger increase in MAP after PMX-F therapy [6]. Furthermore, some studies also demonstrated a decrease in the dose of dopamine or dobutamine after PMX-F [7, 18–20].

Pooled mortality rates were 61.5% in the conventional therapy group and 33.5% in the PMX-F group [6]. In the pooled estimate, PMX appeared to significantly reduce mortality compared with conventional medical therapy (RR: 0.53; 95% CI: 0.43–0.65). The results were similar in both randomized controlled trials (RR: 0.50; 95% CI: 0.37–0.68) and non-randomized controlled trials (RR: 0.55; 95% CI: 0.38–0.81). However, it should be noted that very few of the included studies were planned or powered to specifically assess mortality.

Putting these data into perspective, this systematic review of the published literature found positive effects of PMX-F on blood pressure and dopamine/dobutamine use, the PaO_2/FiO_2 ratio, endotoxin removal, and mortality [6]. It should be noted that many of the analyzed studies were of suboptimal quality, which may have exaggerated the magnitude of these effects.

After the Systematic Review

Following the publication of the meta-analysis, additional studies were published on the use of PMX-F hemoperfusion. A number of these studies confirm the hemodynamic effects described in the meta-analysis, demonstrating either improvement in blood pressure, reduction in vasopressor dose, or both [21–23]. Four of these are briefly discussed below.

In 2007, Cantaluppi et al. [21] investigated the potential of PMX-F hemoperfusion for prevention or attenuation of acute kidney injury related to sepsis. Sixteen patients with Gram-negative sepsis were randomized to receive either standard care alone (Surviving Sepsis Campaign Guidelines) or standard care plus PMX-F. In this small study, cell viability, apoptosis, polarity, morphogenesis and epithelial integrity were evaluated in cultured renal tubular cells and glomerular podocytes incubated with plasma from patients of both groups. They demonstrated that PMX-F therapy reduces the proapoptotic activity of septic plasma on cultured renal tubular cells, via modulation of Fas upregulation, caspase activity and the Bax/Bc12 ratio. Furthermore, loss of plasma-induced polarity and permeability of cell cultures was abrogated with the plasma of patients treated with PMX-F. Proteinuria and urine tubular enzymes were also significantly reduced after PMX-F, but not after standard care alone. These intriguing results pave the way for further study on the role of endotoxin in septic acute kidney injury.

Using immunocytochemical and electron microscopic techniques, Nishibori et al. [24] recently demonstrated the specific removal of activated monocytes from peripheral blood of septic patients by the PMX-F column. PMX-F was able to bind not only free plasmatic lipopolysaccaride, but also monocytes immunoreactive to both CD14 and CD68, cells responsible for the activation of the signaling of Toll-like receptor-4 and the subsequent release of inflammation mediators. They speculated that the removal of such activated monocytes from septic patients may produce beneficial effects by reducing the interaction between monocytes and functionally associated cells, including vascular endothelial cells.

The EUPHAS Study (Early Use of Polymyxin B Hemoperfusion in Abdominal Septic Shock) was also recently published [22] and discussed in further detail by Antonelli et al. [see pp. 83–90]. This study was unique in that it enrolled a targeted patient population which was likely to have high endotoxin levels and in whom definitive source control by surgery was possible. Sixty-four surgical patients with severe abdominal sepsis or septic shock were randomized to either conventional therapy or conventional therapy plus two sessions of PMX-F hemoperfusion. Aside from hemodynamic improvement, PMX-F also appeared to improve organ dysfunction [Sequential Organ Failure Assessment (SOFA) scores] and reduce 28-day mortality (unadjusted HR: 0.43, 95% CI: 0.20–0.94; adjusted HR: 0.36, 95% CI, 0.16–0.80). The authors advocated further multicenter studies to confirm these encouraging findings.

Transplant patients represent an interesting study population, due to their altered immune response and high risk for infections, particularly during the first months after transplantation. The incidence of septic shock in solid organ transplanted patients is reported to be 14%, with a 54% mortality rate [25]. Ruberto et al. [23] investigated the clinical effects of DHP-PMX in solid organ transplant patients who developed severe sepsis or septic shock. Fifteen patients who underwent kidney or liver transplantation and subsequently developed Gram-negative severe sepsis or septic shock were treated by 3 sessions of PMX-F. Similar to the findings in the systematic review, MAP increased from baseline to the third treatment (from 63 ± 5 to 83 ± 4 mm Hg), while the dosage of dobutamine (from 7.5 ± 3 to 3 ± 2 μg/kg/min) and noradrenaline (from 1.3 ± 0.45 to 0.05 ± 0.02 μg/kg/min) were reduced. The PaO_2/FiO_2 ratio also increased (from 234 ± 38.47 to 290 ± 107.48 mm Hg). The authors concluded that the use of PMX-F in association with conventional therapy may be an important aid in solid organ transplant patients with sepsis.

Summary

There is clear biological rationale for endotoxin removal in the clinical management of severe sepsis and septic shock. The current literature seems to provide support for this premise, and provides the basis for further rigorous study. Future trials are planned using assays of endotoxin activity to select and enroll patients.

References

1 Angus D, Wax R: Epidemiology of sepsis: an update. Crit Care Med 2001;29:S109–S116.

2 Manocha S, Feinstein D, Kumar A, Kumar A: Novel therapies for sepsis: antiendotoxin therapies. Expert Opin Investig Drugs 2002;11:1795–1812.

3 Kim JH, Kim SJ, Lee IS, Lee MS, Uematsu S, Akira S, Oh KI: Bacterial endotoxin induces the release of high mobility group box 1 via the IFN-β signaling pathway. J Immunol 2009;182:2458–2466.

4 Shoji H: Extracorporeal endotoxin removal for the treatment of sepsis: endotoxin adsorption cartridge (Toraymyxin) Ther Apher Dial 2003;7:108–114.

5 Cruz DN, Bellomo R, Ronco C: Clinical effects of polymyxin B-immobilized fiber column in septic patients. Contrib Nephrol 2007;156:444–451.

6 Cruz DN, Perazella MA, Bellomo R, de Cal M, Polanco N, Corradi V, Lentini P, Nalesso F, Ueno T, Ranieri VM, Ronco C: Effectiveness of polymyxin B-immobilized fiber column in sepsis: a systematic review. Critical Care 2007;11:R47.

7 Suzuki H, Nemoto H, Nakamoto H, Okada H, Sugahara S, Kanno Y, Moriwaki K: Continuous hemodiafiltration with polymyxin B immobilized fiber is effective in patients with sepsis syndrome and acute renal failure. Ther Apher 2002;6:234–240.

8 Ono S, Tsujinomoto H, Matsumoto A, Ikuta S, Kinoshita M, Michizuki H: Modulation of human leukocyte antigen-DR on monocytes and CD16 on granulocytes in patients with polymyxin B immobilized fiber. Am J Surg 2004;188:150–156.

9 Tani T, Hanasawa K, Kodama M, Imaizumi H, Yonekawa M, Saito M, Ikeda T, Yagi Y, Takayama K, Amano I: Correlation between plasma endotoxin, plasma cytokines, and plasminogen activator inhibitor-1 in septic patients. World J Surg 2001;25:660–668.

10 Nakamura T, Ebihara I, Shoji H, Ushiyama C, Suzuki S, Koide H: Treatment with polymyxin B-immobilized fiber reduces platelet activation in septic shock patients: decrease in plasma levels of soluble P-selectin, platelet factor-4 and betathromboglobulin. Inflamm Res 1999;48:171–175.

11 Ikeda T, Ikeda K, Nagura M, Taniuchi H, Matsushita M, Kiuchi S, Kuroki Y, Suzuki K, Matsuno N: Clinical evaluation of PMX-DHP for hypercytokinemia caused by septic multiple organ failure. Ther Apher Dial 2004;8:293–298.

12 Nakamura T, Kawagoe Y, Matsuda T, Shoji H, Ueda Y, Tamura N, Ebihara I, Koide H: Effect of polymyxin B-immobilized fiber on blood metalloproteinase-9 and tissue inhibitor of metalloproteinase-1 levels in acute respiratory distress syndrome patients. Blood Purif 2004;22:256–260.
13 Kushi H, Miki T, Okamoto K, Nakahara J, Saito T, Tanjoh K: Early haemoperfusion with an immobilized polymyxin B fiber column eliminates humoral mediators and improves pulmonary oxygenation. Critical Care 2005;9:R653–R661.
14 Kushi H, Nakahara J, Miki T, Okamoto K, Saito T, Tanjo K: Hemoperfusion with an immobilized polymyxin B fiber column inhibits activation of vascular endothelial cells. Ther Apher Dial 2005;9:303–307.
15 Nemoto H, Nakamoto H, Okada H, Sugahara S, Moriwaki K, Arai M, Kanno Y, Suzuki H: Newly developed polymyxin B-immobilized fibers improve the survival of patients with sepsis. Blood Purif 2001;19:361–369.
16 Wang Y, Liu Y, Sarker KP, Nakashima M, Serizawa T, Kishida A, Akashi M, Nakata M, Kitajima I, Maruyama I: Polymyxin B binds to anandamide and inhibits its cytotoxic effect. FEBS Lett 2000;470:151–155.
17 Suyama H, Kawasaki Y, Morikawa S, Kaneko K, Yamanoue T: Hiroshima: Early induction of PMX-DHP improves oxygenation in severe sepsis patients with acute lung injury. J Med Sci 2008;57:79–84.
18 Tojimbara T, Sato S, Nakajima I, Fuchinoue S, Akiba T, Teraoka S: Polymyxin B-immobilized fiber hemoperfusion after emergency surgery in patients with chronic renal failure. Ther Apher Dial 2004;8:286–292.
19 Uriu K, Osajima A, Kamochi M, Watanabe H, Aibara K, Kaizu K: Endotoxin removal by direct hemoperfusion with an adsorbent column using polymyxin B-immobilized fiber ameliorates systemic circulatory disturbance in patients with septic shock. Am J Kidney Dis 2002;39:937–947.
20 Kojika M, Sato N, Yaegashi Y, Suzuki Y, Suzuki K, Nakae H, Sigeatu Endo S: Endotoxin adsorption therapy for septic shock using polymyxin B-immobilized fibers (PMX): evaluation by high-sensitivity endotoxin assay and measurement of the cytokine production capacity. Ther Apher Dial 2006;10:12–18.
21 Cantaluppi V, Assenzio B, Pasero D, et al: Polymyxin-B hemoperfusion inactivates circulating proapoptotic factors. Intensive Care Med 2008;34:1638–1645.
22 Cruz DN, Antonelli M, Fumagalli R, et al: Early use of polymyxin B hemoperfusion in abdominal septic shock: the EUPHAS randomized controlled trial. JAMA 2009;301:2445–2452.
23 Ruberto F, Pugliese F, D'Alio A, et al: Clinical effects of direct hemoperfusion using a polymyxin-B immobilized column in solid organ transplanted patients with signs of severe sepsis and septic shock. A pilot study. Int J Artif Organs 2007;30:915–922.
24 Nishibori M, Takahashi HK, Katayama H, et al: Specific removal of monocytes from peripheral blood of septic patients by polymyxin B-immobilized filter column. Acta Med Okayama 2009;63:65–69.
25 Candel FJ, Grima E, Matesanz M, et al: Bacteremia and septic shock after solid-organ transplantation. Transplant Proc 2005;37:4097–4099.

Dinna N. Cruz, MD, MPH
Department of Nephrology, Dialysis and Transplantation, San Bortolo Hospital
Viale Rodolfi 37
IT–36100 Vicenza (Italy)
Tel. +39 0 444993869, Fax +39 0 444993949, E-Mail dinnacruzmd@yahoo.com

Ronco C, Piccinni P, Rosner MH (eds): Endotoxemia and Endotoxin Shock: Disease, Diagnosis and Therapy. Contrib Nephrol. Basel, Karger, 2010, vol 167, pp 83–90

PMX Endotoxin Removal in the Clinical Practice: Results from the EUPHAS Trial

Massimo Antonelli[a] · Roberto Fumagalli[b] · Dinna N. Cruz[c,d] · Nicola Brienza[e] · Francesco Giunta[f] on behalf of the EUPHAS Study Group

[a]Department of Intensive Care and Anesthesiology, Catholic University of Sacred Heart, Rome, [b]Department of Anesthesiology and Intensive Care I, Milano Bicocca University, St Gerardo Hospital, Monza, [c]Department of Nephrology, Dialysis & Transplantation, San Bortolo Hospital, Vicenza, [d]International Renal Research Institute (IRRIV), Vicenza, [e]Department of Emergency and Organ Transplantation, Anesthesia and Intensive Care Unit, University of Bari, Bari, [f]Department of Surgery, University of Pisa, Pisa, Italy

Abstract

Polymyxin B fiber column is a medical device designed to reduce blood endotoxin levels in sepsis. Gram-negative-induced abdominal sepsis is likely to be associated with high circulating endotoxin. In June 2009, the EUPHAS study (Early Use of Polymyxin B Hemoperfusion in Abdominal Sepsis) was published in JAMA. Sixty-four patients who underwent emergency surgery for intra-abdominal infection between December 2004 and December 2007 were enrolled with severe sepsis or septic shock. Intervention patients were randomized to either conventional therapy (n = 30) or conventional therapy plus two sessions of polymyxin B hemoperfusion (n = 34). The main outcome measures were change in mean arterial pressure (MAP) and vasopressor requirement, and secondary outcomes were the PaO_2/FiO_2 (fraction of inspired oxygen) ratio, change in organ dysfunction measured using sequential organ failure assessment (SOFA) scores, and 28-day mortality. At 72 h, MAP increased (76 to 84 mm Hg; p = 0.001) and the vasopressor requirement decreased (inotropic score: 29.9 to 6.8; p = 0.001) in the polymyxin B group, but not in the conventional therapy group (MAP: 74 to 77 mm Hg; p = 0.37; inotropic score: 28.6 to 22.4; p = 0.14). The PaO_2/FiO_2 ratio increased slightly (235 to 264; p = 0.049) in the polymyxin B group, but not in the conventional therapy

The Euphas Study Group are Dinna N. Cruz MD, MPH, Massimo Antonelli MD, Roberto Fumagalli MD, Francesca Foltran MD, Nicola Brienza MD, PhD, Abele Donati MD, Vincenzo Malcangi MD, Flavia Petrini MD, Giada Volta MD, Franco M. Bobbio Pallavicini MD, Federica Rottoli MD, Francesco Giunta MD, Claudio Ronco MD.

group (217 to 228; p = 0.79). SOFA scores improved in the polymyxin B group, but not in the conventional therapy group (change in SOFA: –3.4 vs. –0.1; p = 0.001), and 28-day mortality was 32% (11/34 patients) in the polymyxin B group and 53% (16/30 patients) in the conventional therapy group (unadjusted HR: 0.43, 95% CI: 0.20–0.94; adjusted HR: 0.36, 95% CI: 0.16–0.80). The study demonstrated how polymyxin B hemoperfusion added to conventional therapy significantly improved hemodynamics and organ dysfunction and reduced 28-day mortality in a targeted population with severe sepsis and/or septic shock from intra-abdominal Gram-negative infections.

Endotoxin, one of the principal components on the outer membrane of Gram-negative bacteria, has a key role in the pathogenesis of sepsis and septic shock. High levels of endotoxin activity are associated with worse clinical outcomes [1]. However, the effectiveness of endotoxin-targeted therapy is still controversial [2, 3]. Since the 1970s, polymyxin B (PMX) was discovered to be protective against endotoxin-induced hemodynamic shock, but at the same time was demonstrated to be extremely toxic for the kidney and central nervous system. Since all new anti-endotoxin drugs have failed in demonstrating favorable outcomes in septic shock patients, Toray Industries immobilized PMX to polystyrene fibers in a medical device for hemoperfusion treatment, which can effectively bind endotoxin both in vitro and in vivo, and could potentially interrupt the biological cascade of sepsis [4]. In a recent systematic review by Cruz et al. [5], direct hemoperfusion with the PMX filters appeared to have favorable effects on mean arterial pressure, vasopressor use, the PaO_2/FiO_2 ratio and mortality. However, results were highly questionable due to heterogeneity and methodological limitations of previously available studies.

Septic shock of intra-abdominal origin is often due to Gram-negative or mixed pathogens, and is likely associated with high endotoxin levels. Therefore, it represents a condition in which endotoxin-targeted therapy may be of particular benefit [3]. The EUPHAS study, a randomized controlled trial in a targeted population of patients with septic shock due to intra-abdominal infection after emergency abdominal surgery, was recently published in JAMA on June 17, 2009 [6]. The hypothesis of this study was that PMX hemoperfusion would be associated with improvement of hemodynamic and oxygenation, mitigation of organ dysfunction, and improved survival compared to conventional therapy alone.

Methods

The EUPHAS trial was a prospective multicenter randomized controlled trial conducted in 10 Italian tertiary care intensive care units (ICUs). The patients with severe sepsis/septic shock who had emergency surgery for intra-abdominal infection were randomized to either conventional therapy or conventional therapy plus two sessions of PMX-

Table 1. Physiologic end points at baseline (T0) and at 72 h (T72)

	PMX		p	CONV		p
	T0	T72		T0	T72	
Patients, n	34	34		30	27[a]	
MAP, mm Hg	76 (72–80)	84 (80–88)	0.001	74 (70–78)	77 (72–82)	0.37
Inotropic score	29.9 (20.4–39.4)	6.8 (2.9–10.7)	<0.001	28.6 (16.6–40.7)	22.4 (9.3–35.5)	0.14
VDI (mm Hg^{-1})	4.3 (2.7–5.9)	0.9 (0.3–1.5)	<0.001	4.1 (2.3–6.0)	3.3 (1.3–5.3)	0.26
PaO_2/FiO_2	235 (206–265)	264 (236–292)	0.049	217 (188–247)	228 (199–258)	0.79
Delta renal sofa		–0.3 (–0.7 to 0.1)			0.6 (0.1–1.1)*	

Values are expressed as means (95% CI) * p = 0.01 between groups.
[a] 3 patients died before 72 h.

hemoperfusion. One of the primary outcomes was the improvement in mean arterial pressure (MAP) and vasopressor requirement 72 h after PMX therapy. As indexes of hemodynamic status, the inotropic score (an indicator of both catecholamine and dopamine use) and the vasopressor dependency index (VDI) (an expression of the dose-response relationship between vasopressors and MAP) were used. Outcomes included PaO_2/FiO_2 increase and improvement in organ dysfunction with reductions of the sequential organ failure assessment (SOFA) score at 72 h and in 28-day mortality. Multivariate analysis of mortality end points was done using Cox proportional hazards regression, adjusting for the SOFA score at baseline.

Results

Sixty-four patients (34 in the PMX group and 30 in the conventional therapy group) were enrolled in the trial. There were no significant differences between the two groups at baseline. Both Gram-positive and -negative microorganisms were isolated, and 7 patients had positive cultures from more than one site. Multiple microorganisms were isolated in 34% of the patients.

Physiologic End Points
At 72 h, MAP significantly increased and the inotropic score decreased in the PMX group, but not in the conventional therapy group (table 1). The VDI decreased significantly in the PMX group, but not in the conventional group. There was a borderline significant improvement in PaO_2/FiO_2 in the PMX group, but remained unchanged in the conventional group.

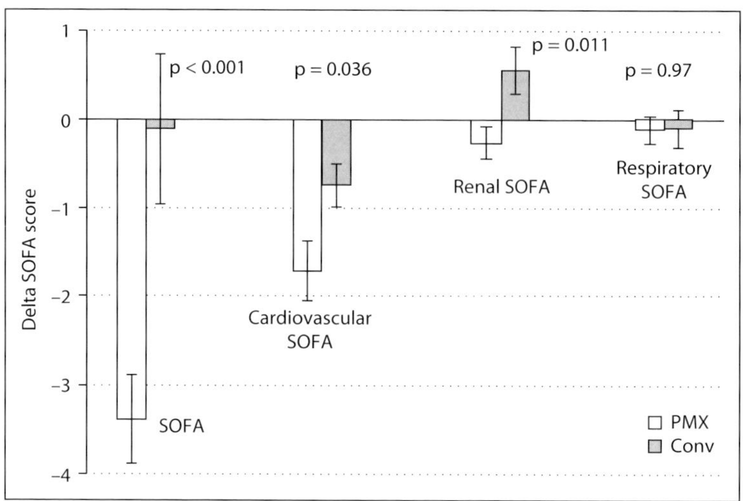

Fig. 1. Change in delta SOFA scores at 72 h; negative values for delta SOFA scores indicate improvement in organ function, while positive values indicate worsening.

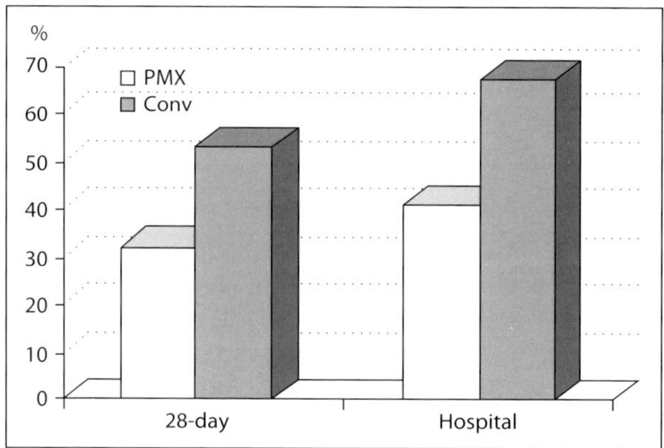

Fig. 2. Crude mortality in the PMX and conventional therapy groups.

The delta SOFA scores showed significant improvement of organ failure in the PMX group (fig. 1). At 72 h, the PMX group had a greater reduction of the total SOFA score and of the cardiovascular and renal components of the SOFA when compared to the conventional group. The two groups were similar in terms of delta respiratory SOFA score.

Mortality End Points

The 28-day mortality was 32% (11/34) in the PMX group and 53% (16/30) in the conventional therapy group (fig. 2). Adjusted for SOFA score, the PMX group had a significant reduction in 28-day mortality (adjusted HR: 0.36, 95% CI: 0.16–0.80, p = 0.012). Hospital mortality was 67% (20/30) in the conventional therapy group, as compared with 41% (14/34) in the PMX group. Adjusted for SOFA score, the PMX group had a significant reduction in hospital mortality rate (adjusted HR: 0.43, 95% CI: 0.21–0.90, p = 0.026).

Discussion

This was the first study focused on a highly selected population of patients with severe sepsis and septic shock due to intra-abdominal sepsis, admitted to the ICU. It strongly reinforces results of previous studies on heterogeneous critical care patients, summarized in the meta-analysis by Cruz et al. [5].

The PMX group showed an adjusted HR of 0.36 and 0.43 for 28-day and all-cause hospital mortality, respectively, when compared with the control group. The delta SOFA scores were significantly better in the PMX group (fig. 1), indicating improvement in overall organ function [7, 8].

The analysis of the delta SOFA showed that its cardiovascular and renal components were mainly responsible for the organ failure improvement recorded after PMX application. The increase of blood pressure and the reduction in vasopressor doses have also been demonstrated by other groups [9, 10]. In the present study, the dose of vasoactive agents (as indicated by the inotropic score) was also reduced, with a significant increase of MAP 72 h after PMX treatment (table 1).

Accordingly, the VDI decreased significantly in the PMX group, but not in the conventional one. Those findings match the results of a previous European pilot study where Vincent et al. [11] showed that patients treated with PMX had significant increases in the cardiac index, left ventricular stroke work index and oxygen delivery index compared with the controls.

The renal component of the SOFA score at 72 h was better in the PMX group (fig. 1), indicating the improvement in the degree of renal organ dysfunction in this group. The proportion of patients treated with renal replacement therapy was similar between the two groups. Earlier studies have demonstrated positive renal effects of PMX therapy [12, 13].

The EUPHAS trial focused on a very homogenous and ill patient population, likely to have high endotoxin levels and in whom definitive source control was possible. Unlike the European pilot study [11], elective abdominal surgery cases were excluded from the EUPHAS trial. Moreover, PMX hemoperfusion sessions were performed twice, rather than once, since the use of two sessions was more consistent with the large Japanese experience with this device.

Caution should be exercised in extrapolating our results to sepsis in a medical population. It is worthwhile to note that while PMX therapy reduces endotoxin levels, and is thus capable of modulating the cascade of events in sepsis, it does not directly address the primary event of sepsis, i.e. the infection. Therefore, PMX cannot afford a definitive cure, but instead could potentially serve as an adjunct to timely and appropriate antibiotic and other medical therapy in severe sepsis.

Larger multicenter studies are needed to confirm these encouraging findings in other patient populations and explore the utility of newer assays for endotoxin activity, both for patient selection and to determine the optimal number of hemoperfusion sessions for individual patients.

The EUPHAS trial presents a series of limitations. First, its highly targeted patient population, although considered one its strengths, contributed to slow patient accrual: only 64 patients were randomized between December 2004 and December 2007. Second, due to the nature of the study intervention, it was not feasible to blind treating physicians to the patient's allocation group, even though the data analysts were blinded. Third, the trial was stopped early based on the results of the interim analysis, following accepted standards for stopping [14–16]. Despite the fact that the sample size was relatively modest, we believe that our results were noteworthy and were in alignment with the results of the meta-analysis on a varied population [5].

Usually small studies tend to overestimate the true magnitude of a clinical effect, but in our trial the benefit with PMX therapy gave a strong biological 'signal'. We feel that the 20% relative reduction in 28-day mortality, as indicated by the higher value of the 95% CI, would be considered clinically relevant in this highly fatal condition. We also acknowledge the inability of the trial to provide definitive answers on the effects of dose duration and number of required PMX treatments. Lastly, as mentioned above, we evaluated the effect of PMX in a surgical population with intra-abdominal sepsis, in whom definitive surgical removal of the septic focus was possible.

In summary, this preliminary randomized controlled trial demonstrated that PMX therapy, when added to conventional medical therapy, was effective in improving 28-day and hospital survival, blood pressure, vasopressor requirement, and degree of organ failure in a targeted population of severe sepsis/septic shock due to intra-abdominal infections.

The adverse events associated with PMX therapy were minimal and similar to those which would be encountered for any extracorporeal therapy in the ICU. Our findings are in agreement with those of other studies in diverse populations, which were summarized in a recent meta-analysis. It is fair to remember that more than one worldwide sepsis expert debated the EUPHAS study [17–20], clearly underlining the study limitations. The author's reply basically accepted the criticism, but stressed the concept that EUPHAS studied a population selected to maximize the signal-noise ratio [21]. As authors, we strongly believe that the EUPHAS trial added a small, but important piece of knowledge to the complex picture of the treatment of sepsis. Our inspiring principle was the concept expressed by Hippocrates 400 years B.C.: 'cure the patient and not the disease'.

References

1 Marshall JC, Foster D, Vincent JL, et al: Diagnostic and prognostic implications of endotoxemia in critical illness: results of the MEDIC study. J Infect Dis 2004;190: 527–534.

2 Opal SM, Glück T: Endotoxin as a drug target. Crit Care Med 2003;31(Suppl 1): S57–S64.

3 Kellum JA: A targeted extracorporeal therapy for endotoxemia: the time has come. Crit Care 2007;11:137.

4 Shoji H, Tani T, Hanasawa K, Kodama M: Extracorporeal endotoxin removal by polymyxin B immobilized fiber cartridge: designing and antiendotoxin efficacy in the clinical application. Ther Apher 1998;2: 3–12.

5 Cruz DN, Perazella MA, Bellomo R, de Cal M, Polanco N, Corradi V, Lentini P, Nalesso F, Ueno T, Ranieri VM, Ronco C: Effectiveness of polymyxin B-immobilized fiber column in sepsis: a systematic review. Crit Care 2007;11:R47.

6 Cruz DN, Antonelli M, Fumagalli R, et al: Early use of polymyxin B hemoperfusion in abdominal septic shock: the EUPHAS randomized controlled trial. JAMA 2009;301:2445–2452.

7 Vincent JL, Moreno R, Takala J, et al: The SOFA (Sepsis-related Organ Failure Assessment) score to describe organ dysfunction/failure: on behalf of the Working Group on Sepsis-Related Problems of the European Society of Intensive Care Medicine. Intensive Care Med 1996;22:707–710.

8 Jones AE, Trzeciak S, Kline JA: The Sequential Organ Failure Assessment score for predicting outcome in patients with severe sepsis and evidence of hypoperfusion at the time of emergency department presentation. Crit Care Med 2009;37:1649–1654.

9 Perego AF, Morabito S, Graziani G, Casella GP, Parodi O: Polymyxin-B direct hemoperfusion (PMX-DHP) in gram negative sepsis (in Italian). G Ital Nefrol 2006;23(Suppl 36):S94–S102.

10 Ruberto F, Pugliese F, D'Alio A, et al: Clinical effects of direct hemoperfusion using a polymyxin-B immobilized column in solid organ transplanted patients with signs of severe sepsis and septic shock: a pilot study. Int J Artif Organs 2007;30:915–922.

11 Vincent JL, Laterre PF, Cohen J, et al: A pilot controlled study of a polymyxin B-immobilized hemoperfusion cartridge in patients with severe sepsis secondary to intra-abdominal infection. Shock 2005;23:400–405.

12 Cantaluppi V, Assenzio B, Pasero D, et al: Polymyxin- B hemoperfusion inactivates circulating proapoptotic factors. Intensive Care Med 2008;34:1638–1645.

13 Nakamura T, Kawagoe Y, Matsuda T, Ueda Y, Koide H: Effects of polymyxin B-immobilized fiber on urinary N-acetyl-B-glucosaminidase in patients with severe sepsis. ASAIO J 2004;50:563–567.

14 Esteban A, Frutos-Vivar F, Ferguson ND, et al: Noninvasive positive-pressure ventilation for respiratory failure after extubation. N Engl J Med 2004;350:2452–2460.

15 Lau WY, Leung TW, Ho SK, et al: Adjuvant intraarterial iodine-131-labelled lipiodol for resectable hepatocellular carcinoma: a prospective randomised trial. Lancet 1999;353:797–801.

16 Pocock SJ: Clinical Trials: A Practical Approach. Chichester, Wiley, 1983.

17 Kellum JA, Uchino S: International differences in the treatment of sepsis: are they justified? JAMA 2009;301:2496–2497.

18 Vincent JL: Polymyxin B hemoperfusion and mortality in abdominal septic shock. JAMA 2009;302:1968.

19 Amaral AC: Polymyxin B hemoperfusion and mortality in abdominal septic shock. JAMA 2009;302:1968–1969.
20 Kida Y: Polymyxin B hemoperfusion and mortality in abdominal septic shock. JAMA 2009;302:1969.
21 Antonelli M, Giunta F, Ronco C: Polymyxin B hemoperfusion and mortality in abdominal septic shock — reply. JAMA 2009;302:1969–1970.

Massimo Antonelli, MD
Department of Intensive Care and Anesthesiology, Catholic University of Sacred Heart
Largo A. Gemelli, 8
IT–00168 Rome (Italy)
Tel. +39 0 630154386, Fax +39 0 63013450, E-Mail m.antonelli@rm.unicatt.it

Ronco C, Piccinni P, Rosner MH (eds): Endotoxemia and Endotoxin Shock: Disease, Diagnosis and Therapy. Contrib Nephrol. Basel, Karger, 2010, vol 167, pp 91–101

Early Management of Endotoxemia Using the Endotoxin Activity Assay and Polymyxin B-Based Hemoperfusion

G. Novelli[a] · G. Ferretti[b] · F. Ruberto[c] · V. Morabito[a] · F. Pugliese[c]

[a]Dipartimento 'P. Stefanini' Chirurgia Generale e Trapianti d'Organo, [b]Dipartimento di Malattie Infettive e Tropicali, [c]Dipartimento di Scienza Anestesiologiche, Medicina Critica e del Dolore, La Sapienza Università di Roma, Rome, Italy

Abstract

Background: We evaluated the ability of the endotoxin activity (EA) assay to determine the need for early intervention for endotoxemia using polymyxin B-based hemoperfusion (PMX-DHP) on septic patients. **Methods:** Twenty-four patients were enrolled. Eleven patients had a high EA level ($\geq$0.6) and were treated with PMX-DHP every 24 h until the EA level was low (<0.4). The remaining 13 patients had EA levels <0.60 and received standard therapy only. **Results:** Two PMX-DHP treatments were performed on 4 patients, three treatments on 6 patients and four treatments on 1 patient. After the therapy, mean arterial pressure increased (69.45 to 84.09 mm Hg; $p < 0.01$), heart rate decreased (111.73 to 77.91 beats/min; $p < 0.01$), white blood cell count decreased (18,380 to 9,550 cells/mm^3; $p < 0.01$), the PMN (polymorphonuclear) percentage decreased (88.45 to 67.82%; $p < 0.01$) and PaO_2/FiO_2 increased (275 to 308.09; $p < 0.01$). All 24 patients survived to the 28-day follow-up. **Conclusion:** The EA assay can identify patients eligible for PMX-DHP treatment and aids its therapeutic dosing.

The pathogenesis of sepsis is driven by a systemic inflammatory response syndrome (SIRS) of the host that involves hemodynamic, respiratory, metabolic and immunologic alterations. The prevalence of SIRS in surgical and ICU patients is very high. One third of these SIRS cases evolve to sepsis, and of these sepsis cases, half evolve to severe sepsis that could be followed by shock in about 25% of cases [1]. The prognosis of these critical patients is related to comorbidities, as well as the severity of the inflammatory response and its development in shock and organ dysfunction or failures. The interval of progression from SIRS to sepsis, severe sepsis and septic shock seems to be inversely correlated to the number of SIRS criteria met by patients [2]. On the other hand, as the severity of the illness shows a graded increase from SIRS to sepsis, severe sepsis and septic shock, the 28-day mortality increases from 10 to 60%. The presence or absence of infections does not influence outcome, although the source of infection does [1].

The speed and appropriateness of diagnosis and therapy administered in the initial hours after the syndrome develops has been shown to influence outcome. Appropriate fluid management and early administration of antibiotics are mandatory during the first hours after a diagnosis of sepsis. Initial antibiotic therapy is often reassessed on the basis of microbiological cultures and clinical data. Moreover, bacterial toxins also play a key role in the onset of SIRS due to their capability to increase capillary permeability and activate neutrophils which induce tissue damage. Microbial components, such as endotoxin [lipopolysaccharide (LPS)], peptidoglycan, peptidoglycan-associated lipoprotein, lipoteichoic acid and other membrane proteins have been reported to activate inflammatory cascades during microbial infections via Toll-like receptors located on the surface of immune cells, i.e. macrophages, dendritic cells or neutrophils [3].

Several studies have reported the effects of endotoxin as a key determinant in the outcome of patients with sepsis; particularly those patients with endotoxemia at ICU admission showed a higher severity of illness and a lower survival than patients without endotoxin detected [4–6].

The primary end point of our study was the early diagnosis of critical endotoxemia in postsurgical abdominal patients showing two or more SIRS criteria in order to identify a subpopulation highly exposed to the progression of the sepsis cascade.

Recently, a new assay was developed that can rapidly detect endotoxin activity (EA) in whole blood. Studies using this assay report EA levels to be significantly correlated with the severity of illness of ICU patients and permits the determination of risk for developing severe sepsis and septic shock. The EA has a predetermined high cutoff level of 0.60 EA units, which is associated with increased risk for adverse outcomes [7].

The relevance of endotoxemia has been confirmed by many studies from Japan and in a recently published Italian multicenter randomized controlled trial (EUPHAS). Those studies evaluated the effectiveness of polymyxin B (PMX)-based hemoperfusion (PMX-DHP) in severe sepsis and septic shock. The treatment was generally administered to patients with suspected Gram-negative infections or endotoxemia (suspected on the basis of an intra-abdominal source). Published data has shown that efficient endotoxin removal has a significant effect on hemodynamics, oxygenation, renal function and survival [8–11]. Thus, clinical experience supports the rationale that endotoxin removal prevents the progression of sepsis to septic shock, multiorgan failure and death.

Therefore, as suggested by recent studies, the use of specific diagnostic tools could identify patients with clinically relevant endotoxemia early and provide useful information to optimize targeted therapeutic interventions.

The secondary end point of our investigation was to evaluate the ability of the EA assay to profitably suggest the therapeutic dosing of selective endotoxin removal on high-risk patients, thus preventing progression of the biological cascade of sepsis.

Materials and Methods

Patient Selection

From April 2008 to April 2009, patients diagnosed for postsurgical sepsis were enrolled in this study. Inclusion criteria were the following findings within the previous 24 h: two or more signs of SIRS such as fever or hypothermia (body temperature >38 or <36°C, respectively), tachycardia (>90 beats/min), tachypnea (>20 breaths/min), an arterial carbon dioxide tension <32 mm Hg, a white blood cell count $>12.0 \times 10^3$/l or $<4.0 \times 10^3$/l or more than 10% immature neutrophils. The degree of organ dysfunction was also evaluated [12, 13]. All the patients received a resuscitation therapy, including fluid challenge and antibiotic therapy.

Endotoxin Activity Assay

EA in whole blood was measured as described by Romaschin et al. [14] using the EAA chemiluminescent assay (Spectral Diagnostics Inc., Toronto, Ont., Canada). EA can be classified as low for EA (<0.4), intermediate for EA (0.4≤ EA <0.6) and high for EA (≥0.6), according to the manufacturer's guidelines.

EA was measured within 24 h of the onset of sepsis symptoms (T0). If EA ≥0.6 was detected (T0 ≡ T1A), patients were then assigned to receive endotoxin removal therapy (PMX-DHP). If the EA level was <0.60, a follow-up EA level was performed at 24 h; if that result remained below the high cutoff, the patient was assigned to receive standard therapy alone.

PMX-DHP Treatment

PMX hemoperfusion was performed using a column of polystyrene fibers to which PMX was covalently bonded [15–17]. Vascular access was obtained with use of double-lumen venous catheters. Each treatment was carried out for 2 h at a flow rate of 100 ml/min.

Patients were treated with PMX-DHP every 24 h if the EA level remained elevated and the patients were exhibiting signs of SIRS. However, PMX-DHP was stopped when an EA level <0.4 was measured. EA levels were measured 1 h after each hemoperfusion in order to evaluate the post-treatment effect, and then again just prior to the start of the next PMX-DHP treatment.

Patients received anticoagulation therapy according to their coagulation profile (platelet count, PTT) at the discretion of the treating physician.

Table 1. Baseline characteristics of patients enrolled in the EA determination

Age	57.8 (range 40–70)
Sex, M/F	10/14
Cause of surgery	Gynecological neoplasias (3) Major abdominal surgery (10) Liver transplantation (7) Kidney transplantation (3) Lung transplantation (1)

Clinical Parameters

Clinical data were reported at T0 (enrolment – T1A), T1B (after the 1st treatment), T2A (before the 2nd treatment), T2B (after the 2nd treatment), T3A (before the 3rd treatment), T3B (after the 3rd treatment), T4A (before the 4th treatment) and T4B (after the 4th treatment).

The following clinical parameters were reported at each time point: body temperature, mean arterial pressure (mm Hg), heart rate (beats/min), the PaO_2/FiO_2 ratio, white blood cell count (cells/mm^3), percentage of neutrophils in the white blood cell count and urine output (ml/day). The comparison of clinical parameters was carried out between the values measured at enrolment (assignment to treatment group) and after the last PMX-DHP session. Episodes of infections were diagnosed by microbiological data and was observed for the period of enrolment up to 72 h.

Statistical Analysis

Data are presented as means (SD) for normally distributed variables or medians (range) for non-normally distributed variables. All data underwent the Wilcoxon signed-ranked test for paired data; statistical significance was obtained for $p < 0.01$.

Results

Twenty-four postsurgical patients showing clinical signs of sepsis were evaluated by the EA assay (Table 1). The endotoxin removal therapy by PMX-DHP was carried out if a high EA (≥0.60 EA units) was found.

Fourteen patients (58%) showed a low or intermediate EA level at the first examination [median 0.32 and range (0.16–0.53)]. These levels did not significantly change after 24 h except for 1 patient. This patient showed an EA of 0.62 after 24 h and was then shifted to the treatment group. Microbiological findings of these patients in the low and intermediate EA level group showed the presence of Gram-positive infections in 7 of 13 patients, 4 infections of mycetes and 2 fungal infections.

Table 2. Characteristics of the treated patients

Cause of surgery	Gynecological neoplasias (2) Major abdominal surgery (4) Liver transplantation (3) Kidney transplantation (2)
Body temperature, °C	39.3±0.3
Mean arterial pressure, mm Hg	69.45±1.86
Heart rate, beats/min	111.73±13.08
PaO_2/FiO_2	275±12.04
White blood cells, cells/mm^3	18,380±2510
Neutrophils, %	88.45±3.72
Platelets, cells/mm^3	208.45±48.47
INR	0.79±0.19
Reported infections	*Pseudomonas aeruginosa* (2) *Klebsiella pneumoniae* (1) Enterobacter (2) *Escherichia coli* (4) *Klebsiella oxytoca* (1) *Proteus mirabilis* (1)

Ten patients (42%) showed a positive EA level at the first examination [median 0.74 (0.62–1.25)]. A total of 11 patients were included in the PMX-DHP group. A summary of the clinical parameters of these patients is reported in table 2. None of the patients were in septic shock and ventilatory support was provided by means of noninvasive techniques.

Two PMX-DHP treatments were performed on 4 patients [median EA = 0.64 (0.62–0.73)], three treatments on 6 patients [median EA = 0.85 (0.74–0.95)] and four treatments on 1 patient (EA = 1.25). The latter patient's EA level was at the upper limit of detection for the EA assay. No adverse events were observed during the 28 treatments performed.

The changes in EA levels measured during each cycle of treatments are presented in figure 1. At the end of the endotoxin removal therapy, the median EA level was 0.29 [range (0.22–0.38)]. Each PMX-DHP session showed a reduction of EA levels of 20–40%. In the patient with a baseline EA of >1.0, the levels were reduced by >20%.

Following the last PMX-DHP treatment, when the EA levels were <0.4, a statistically significant improvement in the hemodynamic parameters was observed.

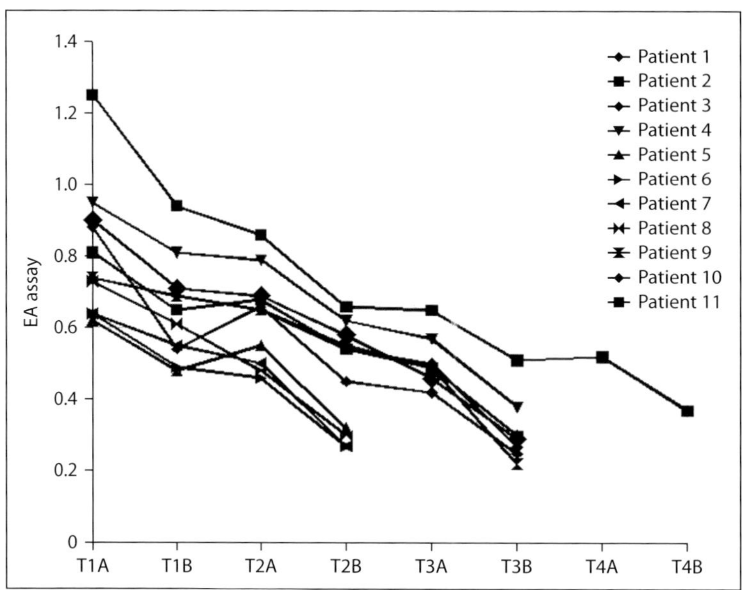

Fig. 1. EA levels of all the patients during the endotoxin removal therapy. Letters A and B identify EA levels prior to and 1 h before each treatment.

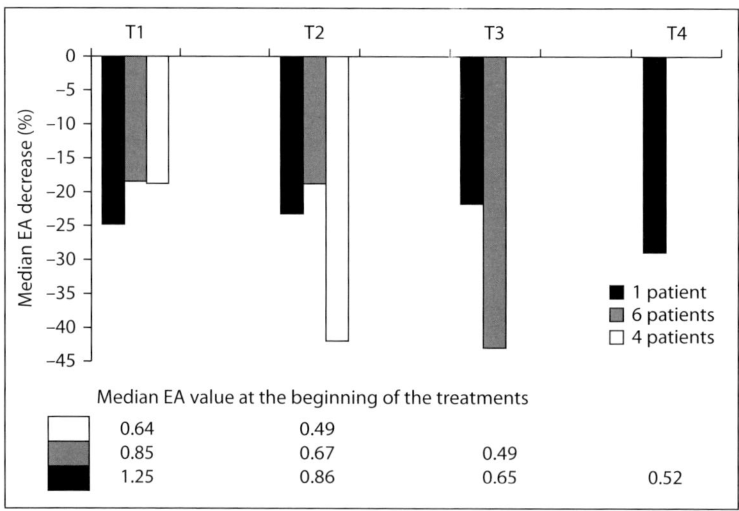

Fig. 2. Percentage reduction of EA levels after each treatment. The grey scales identify patients treated by 2, 3 or 4 hemoperfusions. The median EA values at the beginning of treatments is also reported.

Pre- and post-treatment values of mean arterial pressure and heart rate were observed to have changed from 69.45 ± 1.86 to 84.09 ± 3.75 mm Hg ($p < 0.01$) and from 111.73 ± 13.08 to 77.91 ± 6.59 beats/min ($p < 0.01$), respectively.

The white blood cell count and neutrophil percentage in the white blood cell count respectively changed from 18,380 ± 2,510 cells/mm^3 and 88.45 ± 3.72% to 9,550 ± 1,050 cells/mm^3 and 67.82 ± 8.36% after the completion of therapy ($p < 0.01$). The PaO_2/FiO_2 ratio increased after the PMX-DHP therapy from the initial value of 275 ± 12.04 to 308.09 ± 8.32 ($p < 0.01$).

Satisfactory values of daily urine output were reached at the end of the therapy to a mean daily output of >1,200 ml, except in 1 patient where the daily output was 600 ml, which increased to average 1,000 ml/day after 3 days.

The evaluation of hepatic parameters of the three patients which had undergone a liver transplantation showed an increase in bilirubin (7 ± 2.1 mg/dl), GOT (950 ± 31 mg/dl), GPT (789 ± 55.4 mg/dl) and alkaline phosphatase (356 ± 42.5). All these parameters reached normal values within 5 days after the end of the therapy.

Microbiological findings showed the presence of Gram-negative infections in 9 of 11 patients within 72 h from enrolment. The remaining two patients had negative hemocultures.

All 24 patients survived to the 28-day follow-up.

Discussion

The presence of endotoxemia in patients with severe sepsis and septic shock is well documented in several published studies. Opal et al. [6] in 1999 reported a large series of 253 septic patients where the incidence of endotoxemia was 80%. Patients with marked endotoxemia had a significantly greater 28-day mortality than patients with low values of endotoxemia. However, the natural history of endotoxemia in sepsis is not fully elucidated as there are conflicting reports of the effects of endotoxemia on patients with different ages, comorbidities or degrees of organ dysfunction. This may, in part, be due to variability of the interaction of circulating endotoxin with the host immune system than the absolute level of endotoxin in the bloodstream.

A newly developed method for endotoxin measurement, the EA, incorporates the individual variation associated with the respiratory burst activity of host neutrophils in the presence of an LPS-antibody complex [14, 18]. Marshall et al. [7] in 2004 showed a significant correlation among EA levels and the worsening of clinical parameters, such as the APACHE II score, SOFA score, PaO_2/FiO_2 ratio and white blood cell count. The same study reported a threefold increase of the risk of severe sepsis in patients with high EA levels (EA >0.6) compared to patients with low EA levels (EA <0.4). Moreover, patients with endotoxemia had increased ICU and hospital mortality.

Recently, Valenza et al. [19] evaluated the prevalence of endotoxemia in 102 patients admitted to the ICU after elective surgery. Despite the low level of physiologic derangements at admission (mean APACHE score 8.3), 17% of their patients showed high EA levels. The overall length of stay in the ICU was longer (5.2 days) for patients with high EA levels than patients showing low and intermediate EA levels (1.9 and 1.8 days, respectively). Moreover, patients who underwent abdominal or thoracic procedures exhibited higher EA levels at ICU admission and had a significantly longer ICU length of stay than for those with other surgical procedures.

For patients with septic shock, EA levels have been shown to stratify patients with increasing severity of illness [20]. Although patient numbers were small, septic shock patients with low EA levels had 0% mortality compared to those with intermediate levels (17% mortality) and patients with high EA levels (37.5% mortality).

Taken together, these data suggest that the presence of endotoxin worsens the prognosis of ICU patients and the risk for an adverse outcome is associated with endotoxemia, resulting from or contributing to severe sepsis and septic shock.

In our study, we evaluated the levels of EA in 24 postsurgical patients with sepsis. A high level of EA (>0.6) was found in 42% of these patients at an early stage, i.e. within 24 h of the onset of clinical signs of sepsis. Thus, we hypothesized that an early intervention targeted against endotoxin may improve their predicted sepsis outcome.

PMX-based hemoperfusion has demonstrated the ability to selectively remove endotoxin from the bloodstream and favorably impact patients' outcome. This has been extensively described in Japanese and European literature [9].

Recently, a randomized controlled trial was carried out in a population of patients with septic shock who underwent emergency surgery for intra-abdominal infection. This study showed a greater reduction of the total SOFA score at 72 h in the PMX hemoperfusion group compared to the conventional therapy group (–3.4 vs. –0.1). Cruz et al. [8] also reported a reduction in 28-day mortality in the PMX-DHP-treated group (32 vs. 53% in the conventional group).

Monti et al. [21] carried out a retrospective analysis among patients with septic shock with EA >0.6. Although the study included only small numbers of patients, those assigned to endotoxin removal by PMX-DHP treatment had a significantly shorter length of stay in ICU (21.5 days) compared to patients treated with conventional therapy alone (53.6 days). The difference in mortality of the two groups was clinically relevant but not statistically significant (45% in the conventional therapy group vs. 16% in the PMX-DHP group).

In the current study, postsurgical patients with signs of sepsis and endotoxemia (based on EA values >0.6) were treated with PMX-DHP, specifically targeting EA levels with a therapeutic target to lower levels to <0.4. The 'dosing schedule' (number of PMX-DHP treatments) was determined by measuring the EA levels at 24 h after each treatment and by considering the clinical status of the patient. EA levels were lowered after each treatment. The percentage decrease was higher (approx. 40%) for treatments starting from EA <0.6 than those starting from EA >0.6 (approx. 20%).

This may be explained by characteristics of the assay method and by characteristics of the PMX-DHP method. There is a nonlinear LPS dose response curve for the EA assay [22], and the amount of LPS that can be removed by each PMX-DHP is fixed and determined by the number of available bonding sites of PMX within each cartridge [15].

The antiendotoxin PMX-DHP intervention produced an immediate and significant improvement in hemodynamics and a significant lowering of the inflammatory state was observed as demonstrated by the decrease in white blood cell count and neutrophil percentage. The effect of PMX-DHP on the inflammatory state has been reported by many authors in terms of cytokine production and endothelial cell activation [23–25]. Most recently, Nishibori et al. [26] demonstrated that PMX-DHP selectively removes monocytes from the circulating blood, which may provide a beneficial effect due to a reduced interaction between monocytes and vascular endothelial cells.

In this study, all the patients with elevated endotoxin levels were treated with PMX-DHP and survived to the 28-day follow-up, despite the high risk for bad outcome associated with high EA levels.

This study provides interesting insight into the early management of endotoxemia in sepsis: first, the EA assay showed the ability to identify patients eligible for a targeted therapy against endotoxin; second, the EA assay efficiently measured the effect of the PMX-DHP treatment, thus aiding the therapeutic dosing of this treatment; and third, endotoxemia was also detected in cases without culture evidence of a Gram-negative infection, thereby suggesting the role of translocation from the gastrointestinal tract as a source of endotoxin.

Further studies are needed with a greater number of patients who have more severe disease, and populations at high risk for endotoxemia, such as trauma and pulmonary infections, also need to be selected. Moreover, the EA method may be used to clarify the role of endotoxin translocation in the onset of septic states.

Acknowledgement

We would like to thank Dr. M. Bufi, Dr. A. Morelli and Dr. G. Tritapepe from La Sapienza Università di Roma for allowing us to study some of their patients.

References

1 Brun-Buisson C: The epidemiology of the systemic inflammatory response. Intensive Care Med 2000;26:S64–S74.

2 Rangel-Frausto MS, Pittet D, Costigan M, Hwang T, Davis CS, Wenzel RP: The natural history of the systemic inflammatory response syndrome (SIRS). A prospective study. JAMA 1995;273:117–123.

3 Shimizu T, Endo Y, Tsuchihashi H, Akabori H, Yamamoto H, Tani T: Endotoxin apheresis for sepsis. Transfus Apher Sci 2006;35: 271–282.

4 Danner RL, Elin RJ, Hosseini JM, Wesley RA, Reilly JM, Parillo JE: Endotoxemia in human septic shock. Chest 1991;99:169–175.

5 Hurley JC: Endotoxemia: methods of detection and clinical correlates. Clin Microbiol Rev 1995;8:268–292.

6 Opal SM, Scannon PJ, Vincent JL, White M, Carroll SF, Palardy JE, Parejo NA, Pribble JP, Lemke JH: Relationship between plasma levels of lipopolysaccharide (LPS) and LPS-binding protein in patients with severe sepsis and septic shock. J Infect Dis 1999;180: 1584–1589.

7 Marshall J, Foster D, Vincent J, et al: Diagnostic and prognostic implications of endotoxemia in critical illness: results of the MEDIC study. J Infect Dis 2004;190:527–534.

8 Cruz DN, Antonelli M, Fumagalli R, Foltran F, Brienza N, Donati A, Malcangi V, Petrini F, Volta G, Bobbio Pallavicini FM, Rottoli F, Giunta F, Ronco C: Early use of polymyxin B hemoperfusion in abdominal septic shock: the EUPHAS randomized controlled trial. JAMA 2009;301:2445–2452.

9 Cruz DN, Perazella MA, Bellomo R, de Cal M, Polanco N, Corradi V, Lentini P, Nalesso F, Ueno T, Ranieri VM, Ronco C: Effectiveness of polymyxin B-immobilized fiber column in sepsis: a systematic review. Crit Care 2007;11:R47.

10 Cantaluppi V, Assenzio B, Pasero D, Romanazzi GM, Pacitti A, Lanfranco G, Puntorieri V, Martin EL, Mascia L, Monti G, Casella G, Segoloni GP, Camussi G, Ranieri VM: Polymyxin-B hemoperfusion inactivates circulating proapoptotic factors. Intensive Care Med 2008;34:1638–1645.

11 Ruberto F, Pugliese F, D'Alio A, Martelli S, Bruno K, Marcellino V, Summonti D, Celli P, Perrella S, Cappannoli A, Pietropaoli C, Tosi A, Diana B, Novelli G, Rossi M, Ginanni-Corradini S, Ferretti G, Berloco PB, Pietropaoli P: Clinical effects of direct hemoperfusion using a polymyxin-B immobilized column in solid organ transplanted patients with signs of severe sepsis and septic shock. A pilot study. Int J Artif Organs 2007;30:915–922.

12 Sands KE, Bates DW, Lanken PN, et al: Epidemiology of sepsis syndrome in 8 academic medical centers. JAMA 1997;278: 234–240.

13 Bone RC, Fisher CJ, Clemmer TP, et al: Sepsis syndrome: a valid clinical entity. Crit Care Med 1989;17:389–393.

14 Romaschin AD, Harris DM, Ribeiro MB, et al: A rapid assay of endotoxin in whole blood using autologous neutrophil-dependent chemiluminescence. J Immunol Methods 1998;212:169–185.

15 Shoji H: Extracorporeal endotoxin removal for the treatment of sepsis: endotoxin adsorption cartridge (Toraymyxin). Ther Apher Dial 2003,7:108–114.

16 Vesentini S, Soncini M, Zaupa A, Silvestri V, Fiore GB, Redaelli A: Multi-scale analysis of the Toraymyxin adsorption cartridge. Part I: molecular interaction of Polymyxin-B with endotoxins. Int J Artif Organs 2006;29: 239–250.

17 Fiore GB, Soncini M, Vesentini S, Penati A, Visconti G, Redaelli A: Multi-scale analysis of the Toraymyxin adsorption cartridge. Part II: computational fluid-dynamic study. Int J Artif Organs 2006;29:251–260.

18 Marshall JC, Walker PM, Foster DM, Harris D, Ribeiro M, Paice J, Romaschin AD, Derzko AN: Measurement of endotoxin activity in critically ill patients using whole blood neutrophil dependent chemiluminescence. Critical Care 2002;6:342–348.

19 Valenza F, Fagnani L, Coppola S, Froio S, Sacconi F, Tedesco C, Maffioletti M, Pizzocri MM, Salice V, Ranzi ML, Marenghi C, Gattinoni L: Prevalence of endotoxemia after surgery and its association with ICU length of stay. Critical Care 2009,13:R102.

20 Monti G, Colombo S, Mininni M, Terzi V, Ortisi GM, Vesconi S, Casella G: Why measure endotoxin in septic shock patients? Critical Care 2008;12:S74–S75.

21 Monti G, Terzi V, Mininni M, Colombo S, Vesconi S, Casella G: Polymyxin B hemoperfusion in high endotoxin activity level septic shock patients. Critical Care 2008;12:P458.

22 Foster D, Derzko A, Romaschin A: A novel method for the rapid detection of human endotoxaemia. Clin Lab Int 2004;4:10–12.

23 Kushi H, Miki T, Nakahara J, Okamoto K, Saito T, Tanjoh K: Hemoperfusion with immobilized polymyxin-B column reduces the blood level of neutrophil elastase. Blood Purif 2006;24:212–217.

24 Nakamura T, Matsuda T, Suzuki Y, Shoji H, Koide H: Polymyxin B-immobilized fiber hemoperfusion in patients with sepsis. Dial Transplant 2003;32:602–607.

25 Tani T, Hanasawa K, Kodama M, et al: Correlation between plasma endotoxin, plasma cytokines, and plasminogen activator inhibitor-1 activities in septic patients. World J Surg 2001;25:660–668.

26 Nishibori M, Takahashi HK, Katayama H, Mori S, Saito S, Iwagaki H, Tanaka N, Morita K, Ohtsuka A: Specific removal of monocytes from peripheral blood of septic patients by polymyxin B-immobilized filter column. Acta Medica Okayama 2009;63:65–69.

Prof. Gilnardo Novelli
Dipartimento 'P.Stefanini' Chirurgia Generale e Trapianti d'Organo
Università di Roma 'La Sapienza', Viale del Policlinico, 155
IT–00161 Rome (Italy)
Tel./Fax +39 0649970401, E-Mail novelligilnardo@virgilio.it

Ronco C, Piccinni P, Rosner MH (eds): Endotoxemia and Endotoxin Shock: Disease, Diagnosis and Therapy. Contrib Nephrol. Basel, Karger, 2010, vol 167, pp 102–110

Endotoxin Activity Level and Septic Shock: A Possible Role for Specific Anti-Endotoxin Therapy?

Gianpaola Monti[a] · Maurizio Bottiroli[a] · Giacinto Pizzilli[a] · Maria Minnini[a] · Valeria Terzi[a] · Irene Vecchi[a] · Giovanni Gesu[b] · Paolo Brioschi[a] · Sergio Vesconi[a] · Giampaolo Casella[a]

Departments of [a]Anesthesiology and Intensive Care Medicine, and [b]Microbiology, Niguarda Hospital, Milan, Italy

Abstract

Endotoxin activity (EA) plays an essential role in sepsis syndrome pathogenesis. There has been considerable interest in measuring and removing EA to predict and improve the morbidity and mortality of patients with sepsis. We performed a prospective study to assess the prevalence of EA in critically ill patients and its association with organ dysfunction and outcome, as well as in septic shock. EA (EAA™) was measured within 24 h from onset of refractory septic shock in an intensive care unit. Our study demonstrated that EA level is independent from the type or the source of infection, but reflects the severity of illness in critically ill septic shock patients. Extracorporeal EA removal (PMX-HP) was assessed following our ICU clinical practice. PMX-HP seems to have better outcome, but further studies are required to verify this hypothesis.

Endotoxin or lipopolysaccharide (LPS), an intrinsic component of the Gram-negative outer membrane, is one of the most lethal actors in sepsis syndrome pathogenesis [1]. Although it plays an essential role in the cell wall physiology of the bacterium, endotoxin acts as a potent pattern recognition molecule by alerting, but mostly activating, the innate immune system response of the host at the earliest stages of bacterial invasion [2]. All of these peculiarities underline the 'dual' nature of endotoxin. It is both an 'alarm molecule', warning the host of bacterial invasion within his internal milieu, and a 'trigger molecule' of the pro- and anti-inflammatory cascade in the attempt of orchestrating an efficient antimicrobial defense/elimination. For reasons not entirely clear, in some cases, a 'deregulated or exaggerated' host response to LPS or the release of large quantities of LPS may induce and maintain a sepsis cascade that, if left unchecked, may culminate in multiple organ failure and lethal septic shock [3].

While this 'unique' molecule is a sepsis biomarker, at the same time it could potentially be useful as a diagnostic or a monitoring 'tool', as well as a pivotal mediator in sepsis cascade as an attractive target for therapeutic intervention [4, 5].

Why Measure Endotoxin? Endotoxin as a Sepsis Biomarker

A new and reliable endotoxin activity assay (EAA™), previously described in details, has been recently validated in clinical practice [6]. Briefly EAA™ is based upon the degree of priming of the circulating neutrophil population by endotoxin exposure. At bed side, patient whole blood is incubated with an anti-LPS antibody and then stimulated with opsonized zymosan. The resulting respiratory burst activity of patient white blood cells induced in the presence of endotoxin is detected by chemiluminescence, which is in turn used to quantify the amount of endotoxin activity (EAA™) [7]. EA levels are expressed as units on a scale ranging from 0 to 1: low (EA <0.4 units), intermediate (0.4 < EA <0.6 units) and high (EA >0.6 units).

In a recent trial, Marshall et al. [8] reported that endotoxemia is common in a heterogeneous population of critically-ill patients on the day of admission in ICU: more than 50% of all patients have intermediate or high levels of EA as compared to healthy volunteers. However, only 4% of this global population had a documented Gram-negative infection according to CDC criteria. Moreover, prevalence of Gram-negative infections in the subgroups with intermediate and high levels of EA was not significantly higher, but rather 4.8 and 6.9%, respectively. This discrepancy suggests that (1) EA is not a specific test in detecting Gram-negative infection and, consequently, directing empiric antibiotic therapy before culture tests, and that (2) endotoxemia may derive from sources other than an exogenous Gram-negative infection [9]. The detection of endotoxemia may in fact indicate translocation of viable Gram-negative bacteria or endotoxin from the gastrointestinal tract in the setting of gut barrier dysfunction as observed in hypoperfusion states [10]. In agreement with this, the same study demonstrated that the prevalence of shock was twice as common in critically-ill patients with intermediate and high levels of EA.

There has been considerable interest in the possibility of measuring EA alone or in combination with other markers to predict the outcome of patients with sepsis. Data from the literature states that endotoxemia (or better Limulus lysate amoebocyte-positivity) is found in perhaps 20–40% of septic patients

[9, 11]. Moreover, Casey et al. [12] previously demonstrated that critically ill septic patients with a high LPS-cytokine score had a significantly increased risk of dying compared with patients who had a lower LPS-cytokine score. Unfortunately these results are related to endotoxin measurements by Limulus lysate amoebocyte assay, whose limitations (lack of precision, accuracy and specificity) need to be kept in mind [9, 13].

In the more recent MEDIC study, which was based on EA chemiluminescence assay, there was a significant association between the EA level of critically ill patients and illness severity, as reflected in higher admission APACHE II and SOFA scores in high level EA patients on the day of ICU admission. This subgroup also presented a greater degree of organ dysfunction as reflected by a higher incidence of shock and hypoxemia [8]. Additionally in the same study, increasing levels of EA at the time of admission correlated with increasing ICU mortality. Unfortunately, this study enrolled only a small percentage of patients in sepsis and focused its attention on a heterogeneous population of critically-ill patients, leaving open the question of a possible link between EA and prognosis in sepsis.

On this basis we performed a prospective study to assess the prevalence of endotoxemia in critically ill septic shock patients and its association with outcome. Secondary objectives were to evaluate the relationship between EA and organ dysfunction and prognostic parameters in septic shock. A total of 80 critically ill septic shock patients admitted to our eight-bed general ICU from January 2007 to December 2009 were recruited for this study. EA measurements, assessed by EAA^{TM}, were performed in all ICU patients within 24 h of septic shock onset according to the Surviving Sepsis Campaign (SSC) criteria. All septic shock patients were mechanically ventilated and in vasopressor support, 25% of them were in CRRT (continuous renal replacement therapy) need for anuric renal failure. All patients were treated with standard therapy according to SSC.

When stratified according to EA levels, more than 80% of septic shock patients had intermediate (EA >0.4) or high EA levels (>0.6), confirming the key role of endotoxin as a key trigger of the sepsis cascade. Low EA levels (<0.4) were registered in only 17% of this population. There was no significant difference in the Gram-positive/Gram-negative documented infection ratio between these two groups (data not shown).

More interestingly, our results seem to provide evidence of a good correlation between EA levels and severity of illness in the sepsis context. In fact, 'high EA level septic shock patients' were in need of a significantly higher norepinephrine + epinephrine (NOR+EPI) dose as compared to intermediate and low EA groups (* $p < 0.05$; table 1). Moreover, in our study increasing levels of EA in septic shock patients were associated with increased hospital mortality, even if not statistically different (14, 20 and 32%, respectively, in low, intermediate and high EA septic shock patients). The mortality rate of the high EA group took into account only septic shock patients treated with standard therapy according

Table 1. EA levels and severity of illness in septic shock patients (* $p < 0.05$)

	EAA™ < 0.4	0.4 <EAA™ < 0.6	EAA™ > 0.6
Patients	14 (17%)	25 (31%)	41 (52%)
EAA™ level, units	0.28 ± 0.05	0.49 ± 0.06	0.78 ± 0.14
MAP, mm Hg	79.8 ± 16	79.7 ± 10.2	80.3 ± 13.0
CI	3.99 ± 1.02	3.61 ± 1.01	3.72 ± 1.54
SVRI	1540 ± 343	1686 ± 703	1704 ± 790
NEP+EPI µg/kg/min	0.32 ± 0.23	0.34 ± 0.29	0.67 ± 0.56*
Lactates, mmol/l	4.52 ± 5.71	3.27 ± 2.64	6.15 ± 4.76
SOFA score, points	9.42 ± 4.68	9.68 ± 3.56	11.31 ± 3.79

to SSC (n = 28) and excluded patients treated with antiendotoxin therapy (n = 13), as shown in the following section.

In summary, endotoxemia per se is not able to identify septic shock patients with documented Gram-negative infections. In our study, as previously demonstrated, the presence of high concentrations of circulating endotoxin is apparently independent of the causative microorganism responsible for sepsis (data not shown) [14].

Our results, even if preliminary, support the view that high levels of EA in septic shock patients are correlated with severity of illness, and in particular with hemodynamic dysfunction. In this setting, endotoxemia may be hypothetically seen as the result of splancnic hypoperfusion and may eventually signal the need for improved resuscitation. The role of other possible mechanisms in endotoxemia pathogenesis should be further investigated.

Endotoxemia seems to identify 'a high-risk population' even in septic shock patients, as demonstrated in the trend of increasing mortality associated with increasing EA levels in this population. This observation suggests the hypothesis that the presence of endotoxemia may be able to identify a population of septic shock patients who could probably benefit from specific antiendotoxin therapy.

Endotoxin as a Sepsis Mediator: Target for Antiendotoxin Therapy?

Bacterial endotoxin is one of the most powerful known molecules of bacterial signaling. In human and experimental models, LPS exposure induces a vigorous systemic inflammatory response contributing to generalized inflammation, procoagulant and proapoptotic activity, tissue injury, and septic shock [15].

To better understand the response to endotoxin and the rationale of 'anti-LPS therapy', some considerations should be kept in mind. First, it is the 'highly regulated host response' to LPS, rather than the intrinsic properties of LPS itself, which is responsible for the potentially lethal consequence attributed to this mediator [3, 16]. Second, there is a beneficial homeostatic state between LPS (gastrointestinal flora) and innate immunity. LPS is, in fact, involved in protective immunity supporting the ability to withstand an acute infection challenge. In this view, a 'complete LPS neutralization' may be either beneficial or potentially harmful and useless.

The concept of very early removal of the trigger agent of the sepsis cascade in order to attenuate the excessive activation of innate immune responses and release of host-derived pro/anti-inflammatory mediators was viewed as a reasonable and perhaps optimal therapeutic approach to sepsis. In this context, endotoxemia has been the target of previous clinical trials evaluating the potential benefit of neutralizing or binding endotoxin, aiming at improving the clinical prognosis of patients with presumed Gram-negative infections [17, 18]. Unfortunately, these attempts, which have included the antiendotoxin antibodies HA-1A and E5mAb, have so far failed to show a benefit in this setting [19].

In our opinion, among the numerous approaches for combating endotoxin shock, polymyxin B hemoperfusion (PMX-HP) is at the moment the most attractive. Polymyxin B, an antibiotic with a high affinity for endotoxin, has been bound and immobilized to polystyrene fibers in a medical device for hemoperfusion. As demonstrated by in vitro and in vivo studies, PMX-HP can effectively bind and neutralize endotoxin, reducing the 'lethal' plasmatic endotoxin level [20]. This effect is responsible for the interruption of the sepsis cascade loop, as reflected by the reduced plasma proapoptotic activity in septic shock patients treated with PMX-HP [21].

In a systematic review, Cruz et al. [22] reported that PMX-HP appeared to have a favorable effect on organ dysfunction and mortality in septic shock (mostly Gram-negative). More recently the EUPHASE randomized controlled trial has demonstrated that PMX-HP added to conventional therapy induces a significant improvement in hemodynamic and organ dysfunction with a significant reduction in 28-day mortality in septic shock patients from intra-abdominal Gram-negative infections [23].

Unfortunately, until now no clinical evidence has paired EA in septic shock and specific antiendotoxin therapy (PMX-HP). Considering the higher mortality rate and severity of illness observed in high EA level septic shock patients compared to intermediate and low EA groups, we focused our attention on this group which may 'potentially and hypothetically' benefit from anti-endotoxin therapy. We performed a retrospective analysis on the clinical profile and outcome of two treatment strategies (conventional vs. conventional plus PMX-HP) only in 'high EA level septic shock patients' admitted to our ICU from January 2007 to December 2009.

Table 2. Baseline characteristics of high EA septic shock patients according to strategy of treatment: 'PMX-HP' and 'conventional' groups

	Conventional	PMX-HP	p
Patients, n	28	13	
EAA™ level, units	0.73 ± 0.10	0.89 ± 0.18	0.001
Age, years	58.3 ± 14	59.9 ± 13.7	0.76
MAP, mm Hg	80.7 ± 9.6	79.69 ± 18.7	0.81
SOFA score, points	10.32 ± 4.08	13.23 ± 2.17*	0.02
NEP+EPI μg/kg/min	0.45 ± 0.48	1.15 ± 0.51*	0.001
Lactates, mmol/l	4.42 ± 4.70	7.13 ± 4.50	0.09
PaO_2/FiO_2	255 ± 105	237 ± 118	0.64
CRRT, %	28	23	0.72
ICU length of stay	20.62 ± 18.11	41.1 ± 54.2	0.09

According to our ICU procedures, PMX-HP use has always been restricted to refractory septic shock plus hemodynamic instability (i.e. rapidly increasing dose of vasopressor/inotropes more than 50% from the starting dose in 6 h), three organ failure and/or mechanical ventilation, and full treatment according to SSC. As a result, PMX-HP has been applied as 'adjunctive' therapy to conventional treatment following the decision of the attending physician (based on the criteria mentioned above) and independently from EA level.

We collected data on clinical profile, evolution and outcome from 41 septic shock patients who were stratified to the high EA septic shock group. Twenty-eight of them had been treated with 'conventional' treatment, while thirteen of them were treated with conventional treatment plus PMX-HP, according to our procedures. In the latter group, PMX-HP was performed within 24 h from septic shock diagnosis and in two sessions with an interval of 24 h.

Clinical data was collected at baseline (T0) and at 48 h (T2). There was no significant difference in the Gram-positive/Gram-negative documented infection ratio between the two groups (data not shown). Considering our ICU procedures, the PMX-HP group obviously resulted in a worse clinical manifestation of shock as reflected by a higher need for vasopressors, higher serum lactate and higher SOFA score at baseline (table 2). At T2, despite a higher severity of illness, the PMX-HP treatment group showed a significant and fast reduction in vasopressor requirements ($p < 0.01$; fig. 1), which was not observed in the conventional group. Additionally at T2, a significant improvement of organ dysfunction, reflected by

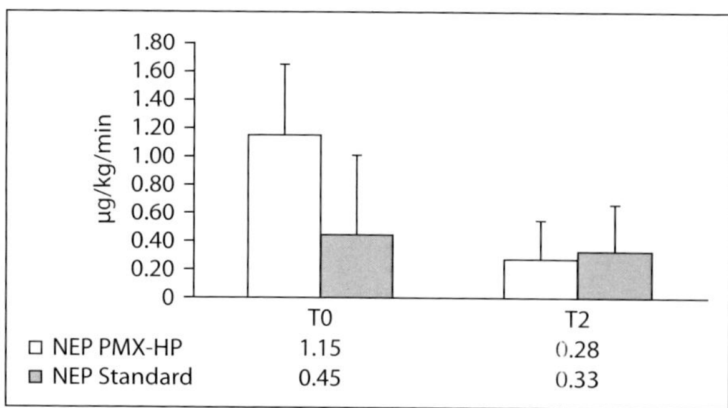

Fig. 1. Evolution of NEP+EPI dose PMX-HP and conventional group of high EA septic shock patients at 48 h. A significant reduction was registered only in the PMX-HP group ($p < 0.001$).

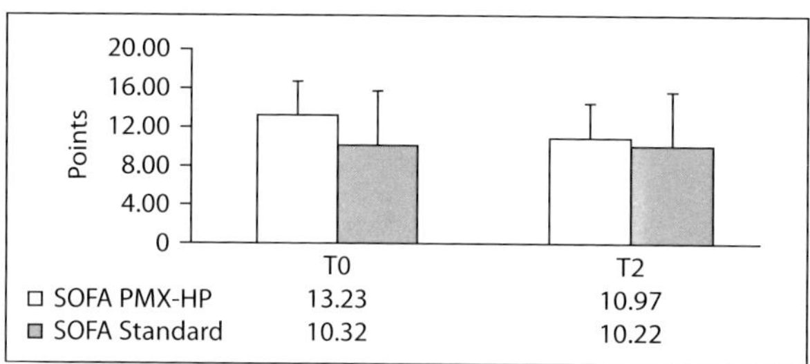

Fig. 2. Evolution of the SOFA score in the PMX-HP and conventional groups of high EA septic shock patients at 48 h. A significant reduction was registered only in the PMX-HP group ($p < 0.05$).

a fall in blood lactate (from 7.13 ± 4.50 to 4.53 ± 4.14 mmol/l, p = 0.001) and SOFA score (13.23 ± 2.17 to 10.97 ± 3.63 points, $p < 0.05$; fig. 2) was observed in the PMX-HP treatment group and not in the conventional group (NS).

The clinical improvement in the PMX-HP group was associated with a reduction of EA level (data not shown; n = 7, due to retrospective nature of the study). The incidence of CRRT at renal dose was similar in both groups, as steroid therapy and APC. No complication related to PMX-HP was recorded. The crude mortality was 15 and 32%, respectively, in the PMX-HP and conventional groups (p = 0.614, χ^2 test). Bearing in mind the limitations of retrospective analysis, a small sample and unbalanced treatments, these clinical results seem to support the extracorporeal removal of endotoxin in refractory septic shock with multiple organ failure. We intentionally did not stress the strong difference in mortality, which could be interesting, but misleading, for any comment in this context.

In summary, our findings are in agreement with data from the literature. PMX-HP added to conventional therapy significantly improved hemodynamic and organ dysfunction and seemed to reduce hospital mortality in a 'targeted' septic shock population. In what was different from the previous studies, this retrospective analysis put the attention on a possible link between EA level and a specific antiendotoxin therapy, PMX-HP. If high EA is, as demonstrated, able to identify a 'high risk septic shock population', we can speculate on the possibility of a targeted intervention aiming at endotoxin removal to treat these patients. With this in mind, our results comparing (independently from the type of infection) two strategies of treatment in 'high EA level septic shock patients' (conventional vs. conventional + PMX-HP) seem to provide evidence of a positive effect of antiendotoxin therapy in terms of fast and significant improvement of organ dysfunction in a severe septic shock population. To our knowledge, this is the first PMX-HP experience deliberately confined to 'extreme septic shock' (baseline NEP + EPI need = 1.15 ± 0.51 μg/kg/min) with multiple organ failure. Larger and prospective randomized studies are indicated to confirm these encouraging findings in a less severe population. The encouraging data related to EA level reduction post-PMX-HP treatment need to be validated by an analysis of the EA level kinetic in all high EA septic shock patients.

Conclusions

In accordance with data from the literature, our study demonstrated that EA level is independent from the type or the source of infection, but reflects the severity of illness in critically ill septic shock patients. EA level may be used as a marker to stratify a 'high risk septic shock population' or inversely to lay out a favorable evolution of sepsis. The identification of a 'high-risk' septic shock population by high EA level can open two original scenarios:

- Could the diagnosis of septic shock turn into 'with or without high EAA™ level' in the future? Furthermore, should endotoxin testing be part of routine clinical assessment in septic shock patients?
- Is there any place for endotoxin removal in septic shock with high EA level? Would it be dependent or independent from the type of infection?

Further study is required before we can accept either of these suggestions.

References

1 Ulevitch RJ, Tobias PS: Recognition of Gram-negative bacteria and endotoxin by innate immune system. Curr Opin Immunol 1999;11:19–22.

2 Rietschel ET, Brade H, Holst O, et al: Bacterial endotoxin: chemical constitution, biological recognition, host response and immunological detoxification. Curr Top Microbiol Immunol 1996;216:39–81.

3 Opal SM: The host response to endotoxin, anti-LPS strategies and the management of severe sepsis. Int J Med Microbiol 2007;297: 365–377.

4 Balk RA: Endotoxemia in critically ill patients: why a reliable test could be beneficial. Crit Care 2002;6:289–290.

5 Klein DJ, Derzko A, Seely A, Foster D, Marshall J: Marker or mediator? Changes in endotoxin activity as a predictor of adverse outcomes in critical illness. Crit Care 2005;9(Suppl 1):161–165.
6 Marshall JC, Walker PM, Foster DM, et al: Measurement of endotoxin activity in critically ill patients using whole blood neutrophil dependent chemiluminescence. Crit Care 2002;6:342–348.
7 Romaskin AD, Harris DM, Ribeiro MB, et al: Rapid assay of endotoxin in whole blood using autologous neutrophil-dependent chemiluminescence. J Immunol Methods 1998;212:169–185.
8 Marshall JC, Forster D, Vincent JL, et al: Diagnostic and prognostic implications of endotoxemia in critical illness: results of the MEDIC study. J Infect Dis 2004;190:527–534.
9 Cohen J: The detection and interpretation of endotoxemia. Intensive Care Med 2000;26(Suppl 1):s51–s56.
10 Fink MP, Aranow JS: Gut barrier dysfunction and sepsis; in Fein A(ed): Sepsis and Multiorgan Failure. Baltimore, William & Wilkins, 1997, pp 383–407.
11 Guidet B, Barakett V, Vassal T, Petit JC, Offenstadt G: Endotoxemia and bacteremia in patients with sepsis syndrome in the intensive care unit. Chest 1994;106:1194–1201.
12 Casey LC, Balk RA, Bone RC: Plasma cytokine and endotoxin levels correlate with survival in patients with the sepsis syndrome. Ann Intern Med 1993;119:771–778.
13 Cohen J, McConnell JS: Observations on the measurement and evaluation of endotoxemia by a quantitative Limulus lysate microassay. J Infect Dis 1984;150:916–924.
14 Opal SM, Scannon PJ, Vincent JL, et al: Relationship between plasma levels of lipopolysaccharide (LPS) and LPS-binding protein in patients with severe sepsis and septic shock. J Infect Dis 1999;180:1584–1589.
15 Zhao B, Bowden RAS, Stavchansky SA, Bowmann PD: Human endothelial cell response to Gram-negative LPS assessed with cDNA microarrays. Am J Physiol Cell Physiol 2001;281:C1587–C1595.
16 van der Poll T, Opal SM: Host-pathogen interactions in sepsis. Lancet Infect Dis 2008; 8:32–43.
17 Opal SM, Glück T: Endotoxin as a drug target. Crit Care Med 2003;31(Suppl):57–64.
18 Kellum JA: A targeted extracorporeal therapy for endotoxemia: the time has come. Crit Care 2007;11:137.
19 McCloskey RV, Straube RC, Sanders C, Smith SM, Smith CR: Treatment of septic shock with human monoclonal antibody HA-1A. A randomized, double-blind, placebo-controlled trial. CHESS Trial Study Group. Ann Intern Med 1994;121:1–5.
20 Shoji H: Extracorporeal endotoxin removal for the treatment of sepsis: endotoxin adsorption cartridge (Toraymyxin). Ther Apher Dial 2003;7:108–114.
21 Cantaluppi V, Assenzio B, Pasero D, et al: Polymyxin B-hemoperfusion inactivates circulating proapoptotic factors. Intensive Care Med 2008;34:1638–1645.
22 Cruz DN, Perazella MA, Bellomo R, et al: Effectiveness of polymyxin B-immobilized fiber column in sepsis: a systematic rewiew. Crit Care 2007;11:R47.
23 Cruz DN, Antonelli M, Fumagalli R, et al: Early use of polymyxin B-hemoperfusion in abdominal septic shock. JAMA 2009;301: 2445–2452.

Gianpaola Monti, MD
Intensive Care Unit, Niguarda Hospital
Piazza Ospedale Maggiore 3
IT–20100 Milano (Italy)
E-Mail Gianpaola.monti@ospedaleniguarda.it

Ronco C, Piccinni P, Rosner MH (eds): Endotoxemia and Endotoxin Shock: Disease, Diagnosis and Therapy. Contrib Nephrol. Basel, Karger, 2010, vol 167, pp 111–118

Endotoxin Removal: How Far from the Evidence? From EUPHAS to EUPHRATES

Jean-Sebastien Rachoin[a] · Debra Foster[b] · R. Phillip Dellinger[a]

[a]Robert Wood Johnson Medical School, University of Medicine and Dentistry of New Jersey, Department of Medicine, Division of Critical Care Medicine, Cooper University Hospital, Camden, N.J., USA; [b]Spectral Diagnostics Inc., Toronto, Ont., Canada

Abstract

There is a large amount of support for the safety of polymyxin-B (PMX-B) hemoperfusion in the treatment of septic shock from Japan and Europe. There is also support for potential efficacy, although randomized controlled trials are few and conflicting. PMX-B hemoperfusion represents a promising new treatment that could significantly improve survival. Previous clinical trials of PMX-B have been criticized for methodological issues, such as the absence of blinding, the use of surrogate outcomes and lack of longer term mortality outcomes. The variability in the number of treatment cartridges used, the selection of subjects based on likelihood of endotoxin presence without endotoxin measurement, and small sample sizes in mainly single-center trials have also been cited. The newly designed EUPHRATES trial (Evaluating Use of Polymyxin Hemoperfusion in a Randomized Controlled Trial of Adults treated for Endotoxemia and Septic Shock) addresses many of the methodological issues and represents a significant opportunity to test for clinical efficacy of endotoxin removal in the critically ill septic patient.

With more than 750,000 annual cases in the United States and 200,000 associated deaths, sepsis represents a major public health concern [1]. Although mortality attributable to septic shock has decreased over the last two decades, it remains the most common cause of death in the intensive care unit [2]. The U.S.-based Surviving Sepsis Campaign (SSC) has made substantial progress in improving the process of care in patients with sepsis through evidence-based guidelines and, more importantly, as an associated formal performance improvement program (SSC sepsis bundles). Recently published data indicate that implementation of the SSC PI program was associated with a significant decrease in mortality [3]. Nevertheless, mortality remains unacceptably high. The average cost to treat a patient with severe sepsis is estimated to be at least

USD 50,000 with an annualized total cost of USD 17 billion in the United States [1]. Therefore, new effective treatments targeting patients most likely to respond has important potential to improve clinical outcomes and reduce healthcare costs.

Treating Sepsis: Importance of Targeting Endotoxin

Endotoxin, a lipopolysaccharide from the cell wall of Gram-negative bacteria that stimulates immune response by binding to the Toll-like receptor 4, has been well characterized as one of the most potent triggers for the sepsis inflammatory cascade [4].

Healthy humans given intravenous endotoxin consistently experience a systemic inflammatory response that can result in myocardial depression, lower blood pressure, and if sustained, multiple organ failure and death [5, 6]. In critically ill patients, elevated levels of endotoxin are known to be associated with increased risk of death [7, 8].

Up to 80% of patients with severe sepsis have elevated levels of endotoxin, while at least 30% have very high levels [7]. Elevated levels of endotoxin can occur in the setting of a culture-documented Gram-negative infection, as well as in the setting of a Gram-positive infection, fungal infection or in other cases of septic shock where no microbiologic source is identified [7, 9, 10]. In such patients, the source of endotoxin is thought to be related to the movement of endotoxin across the gut mucosal barrier in the setting of shock, hypoxemia and gut hypoperfusion [4, 11].

Targeting endotoxemia with a directed therapy in those patients known to have high levels is an attractive therapeutic approach to decrease mortality in this patient population in the United States [12]. A rapid turn around assay for measuring endotoxin is available for use in the U.S. [7]. In Japan, a polymyxin B (PMX-B) embedded hemoperfusion cannister is available for removal of endotoxin. The potential to marry these two technologies to test the effect of endotoxin removal in patients with septic shock and known high levels of endotoxin exists.

The Use of PMX-B in Japan

PMX-B is a potent antibiotic that binds the lipid A portion of endotoxin [13]. Intravenous use is associated with risk of neurotoxicity and nephrotoxicity [14]. These adverse effects do not occur when PMX-B is immobilized and acts as a ligand in an extracorporeal adsorbent hemoperfusion column, where it binds to endotoxin and removes it from the bloodstream [15].

Since the first published report of PMX-B clinical use in 1994 [16], there have been over 50 articles published reporting PMX-B use in over 1,400 critically ill patients. It has been approved for use in Japan since 1993 and in Europe since 1998. More than 70,000 patients have been treated with PMX-B in Japan and Italy over the last 15 years.

PMX-B has been shown to decrease levels of endotoxin in patients with septic shock. In addition, endotoxin removal with PMX-B has been shown to modulate the inflammatory response by decreasing levels of cytokines such as TNF and IL-6 in human studies [17].

In 2003, Nakamura et al. [18] published the result of the largest study to date conducted on PMX-B. In this open-labeled controlled study, they enrolled 314 patients with severe sepsis. 206 patients met the criteria for use of PMX-B based on a culture-confirmed infection or elevated endotoxin levels and the failure of one organ. The PMX group received two hemoperfusion treatments in 24 h. Mortality at 28 days for the PMX-treated group was 32%, in standard care it was 67% ($p < 0.01$). Following PMX-B use, survivors had significant improvements in blood pressure, heart rate, body temperature and the PO_2/FiO_2 ratio ($p < 0.05$). In addition, survivors treated with PMX-B had significant reductions in the levels of endotoxin as well as inflammatory mediators: IL-6, TNF, endothelin-1, soluble IL-2 receptors, and PF-4 ($p < 0.05$ for all individual mediators).

In a single center trial in Japan, Nemoto et al. [19] reported on 98 patients randomized to PMX-B (n = 54) or conventional therapy (n = 44). The overall 28-day survival rate for PMX-B-treated patients was 41% compared to 11% in the conventional therapy group ($p < 0.002$). Mean APACHE II scores were not significantly different between the groups; however, a subgroup analysis of mortality based on an APACHE II cutoff of 30 points showed that for patients with APACHE II scores >30, PMX-B treatment did not significantly improve survival (7 vs. 0%; p = 0.59). However, in this trial, PMX-B use significantly reduced endotoxin levels in all patients from 38.6 ± 5.7 to 21.4 ± 2.0 pg/ml (p = 0.006) and PMX-B significantly improved mean blood pressure for APACHE II score groups <30. There was a trend toward improved outcomes in those patients with an APACHE II score <30 versus those with >30, suggesting that patients with severely advanced disease may not benefit. This study is limited by small numbers and numerous confounders.

Cruz et al. [17] conducted a systematic review of trials that studied the effects of PMX-B hemoperfusion in patients with sepsis. The meta-analysis included 28 publications from 1998 to 2006 with a pooled sample size of 1,425 patients, 978 of whom received PMX-B and 447 standard medical care alone. The major finding was that PMX-B therapy was associated with significantly lower hospital mortality (61.5% in the standard medical care group vs. 33.5% in the PMX-B group, relative risk: 0.53, 95% CI: 0.43–0.65, $p < 0.001$). This dramatic reduction in mortality was accompanied by significant hemodynamic improvement such as an increase in mean arterial pressure of 19 mm Hg (95% CI: 15–22, $p < 0.001$) and a decrease in vasopressor dose by 1.8 µg/kg/min (95% CI: 0.4–3.3, p = 0.01) after PMX use. In addition, the mean PO_2/FiO_2 ratio increased by 32 units (95% CI: 23–41, p < 0.001). The pooled estimate showed that endotoxin levels decreased by 21.2 pg/ml (95% CI: 17.5–24.9) after PMX-B, representing a decrease of 33–80% from pre-PMX levels. The conclusion reached from this review of the published literature is that PMX-B appears to have favorable effects on critical physiological, as well as mortality, outcomes.

The European Experience

European Pilot Study of PMX in Sepsis with Abdominal Infection

A European multicenter randomized trial for PMX-B use in intra-abdominal sepsis demonstrated an improvement in physiologic measurements, but did not show a mortality benefit [20]. Seventeen (out of a total of 36) postsurgical patients with intra-abdominal sepsis were randomized to PMX-B. In this study, the use of the PMX-B was limited to a single cartridge only. Endotoxin levels were not used as an inclusion criterion. This trial did not show a significant reduction in endotoxin levels using the Limulus amoebocyte lysate assay. Not all patients had elevated levels of endotoxin at baseline and endotoxin levels were highly variable between the groups at all time points.

For patients who received the PMX cartridge, there was an increase in mean arterial pressure of 6 ± 13.6 mm Hg from baseline to day 2 (p = 0.006). Mortality was 28% (5/18) in control patients, and 29% (5/17) in the PMX-treated group.

The author's conclusions were that the PMX-B was safe, with no serious adverse events noted, and its use is associated with improved hemodynamic status and cardiac function. A limitation of this study may have been that one cartridge hemoperfusion run was used, which may have been inadequate to optimize endotoxin removal. By not prescreening for endotoxin levels, a number of patients who were treated with PMX-B did not have high endotoxin levels.

The Early Use of Polymyxin B Hemoperfusion in Abdominal Sepsis – EUPHAS Trial

A recently published multicenter randomized clinical trial (ClinicalTrials.gov with identifier: NCT00629382) showed that PMX-B, when added to conventional medical therapy for 64 patients with severe sepsis and septic shock from intra-abdominal infections, resulted in a clinically significant reduction in 28-day mortality [32% PMX group versus 53% in the conventional group, (unadjusted HR: 0.42, 95% CI: 0.20–0.94; adjusted HR: 0.36, 95% CI: 0.16–0.80; p = 0.012)] [21]. In addition, at 72 h post-treatment, the patients treated with PMX-B had a statistically significant improvement in a composite organ failure score [SOFA delta –3.4 (95% CI: –4.4 to –2.4)] compared to patients receiving conventional therapy alone [–0.1 (95% CI: –1.7 to 1.5); p < 0.001]. Patients treated with PMX-B had a clinically and statistically significant increase in mean arterial pressure (76–84 mm Hg, p = 0.001) and the vasopressor requirement was significantly decreased.

This group of patients was chosen as the most likely to have high endotoxin levels based on an intra-abdominal source of sepsis. However no endotoxin levels were measured.

The initial sample size proposed for the study was 120 patients (60 treatment/60 control).

The study enrolled 64 patients between December 2004 and December 2007 in 10 hospitals in Italy, but was stopped early based on results from one planned interim analysis after 30 patients had been enrolled in each group and followed to hospital discharge. The president of the host hospital's ethics committee for the lead investigative site declared it would be unethical to continue the study based on the potential benefit for a group of patients that have a high hospital mortality risk. There has been criticism of this decision.

The 28-day mortality rate was 32% (11/34) for the PMX-B group and 53% in the conventional group (16/30). Following adjustment for the SOFA score, the PMX group had a significant reduction in 28-day mortality (adjusted HR: 0.36, 95% CI: 0.16–0.80, p = 0.012). The mortality benefit continued beyond 28 days through to hospital discharge.

This was the largest study conducted outside Japan for PMX-B use in patients with septic shock. For patients who received PMX, there was a 21% absolute risk reduction for hospital mortality (relative risk reduction: 39.6%).These findings were consistent with other studies in diverse populations as summarized by Cruz et al. [17] in the meta-analysis mentioned above.

Lessons Learned from Previous Trials and Future Directions

The EUPHRATES trial (Evaluating the Use of Polymyxin B Hemoperfusion in a Randomized Controlled Trial of Adults Treated for Endotoxemia and Septic Shock) was designed to test the safety and efficacy of PMX-B in a robust randomized controlled trial in order to address previous criticisms of the published literature that supported the clinical utility of PMX-B hemoperfusion in septic shock.

One major methodological criticism of previous trials has been lack of blinding. As all studies were open-label, there was the risk of introducing a bias which could have 'artificially' prolonged survival in treated patients. Blinding in the past has been a challenge due to the ethical problems of using a sham control. However, the EUPHRATES trial will attempt to blind the study for those in charge of management decisions using an innovative procedure to blind assessors without the use of a sham control. Another important criticism of EUPHAS and other previous trials of PMX-B is the short-term end point for mortality or other surrogate end points used. In more than 50 published studies, only two had a mortality end point greater than 30 days [22–23] (60 days). Most, in fact, used surrogate end points, including hemodynamic improvement and organ failure improvement. EUPHRATES, although powered for 28-day mortality as the primary outcome, will also track mortality for up to 1 year.

There has been inconsistency in previous trials over the number of columns used to treat patients. While the bulk of preclinical data and data from Japanese studies supports two treatments, the European pilot study from Vincent et al. [20] only used one. The EUPHRATES protocol will use two treatments per patient as was done in EUPHAS.

One of the clinical challenges preventing the verification of successful anti-endotoxin strategies has been ensuring that subjects enrolled in trials targeting neutralization or removal of endotoxin have endotoxemia [24]. No endotoxin measurements were included in any of the previous studies, either as inclusion or monitoring criteria, as the assay was not widely available at the time the trials were conducted. EUPHAS and the European pilot study were designed to enrich the study population with patients most likely to be endotoxemic by enrolling patients with intra-abdominal postsurgical sepsis and a high suspicion for Gram-negative infections. However, best clinical efforts did not result in a homogenous population with respect to endotoxemia and as shown in the European pilot study, where mortality benefit may have been hampered by including patients without endotoxemia.

In 2003, the FDA cleared the Endotoxin Activity Assay (EAA™), K021885. Prior to this, there was no FDA-approved method of measuring endotoxin in patients. This assay can now be used to identify high levels of endotoxin in patients with sepsis. A large multicenter observational study of endotoxemia in critical illness [7] used the EAA™ and reported that high levels of endotoxin were found in patients with confirmed infections from Gram-negative or Gram-positive organisms, as well as culture-negative patients, the latter presumably due to get translocation of endotoxin.

As PMX-B is being tested to improve outcomes by removing endotoxin from the bloodstream, patients with no or low endotoxin levels will not be enrolled in the EUPHRATES study. In order to ensure the appropriate selection of patients for the EUPHRATES trial, a two-step eligibility process is planned. All patients that meet the prerequired criteria for septic shock will be screened for endotoxemia using the EAA™. Only patients with high endotoxin levels will be eligible for randomization.

Finally to overcome the criticism of trials that enrolled smaller numbers of patients in single centers, the EUPHRATES trial will be the largest multicentered trial for an extracorporeal sepsis therapy that has been conducted.

Conclusion

PMX-B hemoperfusion is a promising treatment for patients with septic shock and documented high levels of endotoxin. Previous clinical trials support the potential for significant clinical benefit, but were hampered by issues such as the absence of blinding, surrogate and short-term outcomes, the variable number of treatments used, and the lack of endotoxin measurement prior to randomization. The EUPHRATES trial will address those specific issues and is expected to begin enrolling patients in the second or third quarter of 2010.

References

1 Martin GS, Mannino DM, Eaton S, Moss M: The epidemiology of sepsis in the United States from 1979 through 2000. N Engl J Med 2003;348:1546–1554.

2 Angus DC, Linde-Zwirble WT, Lidicker J, Clermont G, Carcillo J, Pinsky MR: Epidemiology of severe sepsis in the United States: analysis of incidence, outcome, and associated costs of care. Crit Care Med 2001;29:1303–1310.

3 Castellanos-Ortega A, Suberviola B, Garcia-Astudillo LA, et al: Impact of the surviving sepsis campaign protocols on hospital length of stay and mortality in septic shock patients: results of a 3-year follow-up quasi-experimental study. Crit Care Med 2010, E-pub ahead of print.

4 Cinel I, Dellinger RP: Advances in pathogenesis and management of sepsis. Curr Opin Infect Dis 2007;20:345–352.

5 Taveira da Silva AM, Kaulbach HC, Chuidian FS, Lambert DR, Suffredini AF, Danner RL: Brief report: shock and multiple-organ dysfunction after self-administration of Salmonella endotoxin. N Engl J Med 1993;328:1457–1460.

6 Suffredini AF, Fromm RE, Parker MM, et al: The cardiovascular response of normal humans to the administration of endotoxin. N Engl J Med 1989;321:280–287.

7 Marshall JC, Foster D, Vincent JL, et al: Diagnostic and prognostic implications of endotoxemia in critical illness: results of the MEDIC study. J Infect Dis 2004;190:527–534.

8 Casey LC, Balk RA, Bone RC: Plasma cytokine and endotoxin levels correlate with survival in patients with the sepsis syndrome. Ann Intern Med 1993;119:771–778.

9 Opal SM, Gluck T: Endotoxin as a drug target. Crit Care Med 2003;31(Suppl 1): S57–S64.

10 Danner RL, Elin RJ, Hosseini JM, Wesley RA, Reilly JM, Parillo JE: Endotoxemia in human septic shock. Chest 1991;99:169–175.

11 Clark JA, Coopersmith CM: Intestinal crosstalk: a new paradigm for understanding the gut as the 'motor' of critical illness. Shock 2007;28:384–393.

12 Kellum JA: A targeted extracorporeal therapy for endotoxemia: the time has come. Crit Care 2007;11:137.

13 Bhor VM, Thomas CJ, Surolia N, Surolia A: Polymyxin B: an ode to an old antidote for endotoxic shock. Mol Biosyst 2005;1: 213–222.

14 Danner RL, Joiner KA, Rubin M, et al: Purification, toxicity, and antiendotoxin activity of polymyxin B nonapeptide. Antimicrob Agents Chemother 1989;33: 1428–1434.

15 Shoji H: Extracorporeal endotoxin removal for the treatment of sepsis: endotoxin adsorption cartridge (Toraymyxin). Ther Apher Dial 2003;7:108–114.

16 Aoki H, Kodama M, Tani T, Hanasawa K: Treatment of sepsis by extracorporeal elimination of endotoxin using polymyxin B-immobilized fiber. Am J Surg 1994;167:412–417.

17 Cruz DN, Perazella MA, Bellomo R, et al: Effectiveness of polymyxin B-immobilized fiber column in sepsis: a systematic review. Crit Care. 2007;11:R47.

18 Nakamura T, Matsuda T, Suzuki Y, Shoji H, Koide H: Polymyxin B-immobilized fiber hemoperfusion in patients with sepsis. Dial Transplant 2003;32:602–607.

19 Nemoto H, Nakamoto H, Okada H, et al: Newly developed immobilized polymyxin B fibers improve the survival of patients with sepsis. Blood Purif 2001;19:361–368, discussion 368–369.

20 Vincent JL, Laterre PF, Cohen J, et al: A pilot-controlled study of a polymyxin B-immobilized hemoperfusion cartridge in patients with severe sepsis secondary to intra-abdominal infection. Shock 2005;23:400–405.
21 Cruz DN, Antonelli M, Fumagalli R, et al: Early use of polymyxin B hemoperfusion in abdominal septic shock: the EUPHAS randomized controlled trial. JAMA 2009;301:2445–2452.
22 Nakamura T, Ushiyama C, Suzuki Y, et al: Combination therapy with polymyxin B-immobilized fibre haemoperfusion and teicoplanin for sepsis due to methicillin-resistant *Staphylococcus aureus*. J Hosp Infect 2003;53:58–63.
23 Nakamura T, Ushiyama C, Suzuki Y, et al: Hemoperfusion with polymyxin B-immobilized fiber in septic patients with methicillin-resistant *Staphylococcus aureus*-associated glomerulonephritis. Nephron Clin Pract 2003;94:c33–c39.
24 Marshall JC: Sepsis: rethinking the approach to clinical research. J Leukoc Biol 2008;83:471–482.

R. Phillip Dellinger, MD
Head, Division of Critical Care Medicine, Cooper University Hospital
One Cooper Plaza, 393 Dorrance
Camden, NJ 08103 (USA)
Tel. +1 856 342 2567, Fax +1 856 968 8306, E-Mail dellinger-phil@cooperhealth.edu

Ronco C, Piccinni P, Rosner MH (eds): Endotoxemia and Endotoxin Shock: Disease, Diagnosis and Therapy. Contrib Nephrol. Basel, Karger, 2010, vol 167, pp 119–125

Endotoxin Removal: How Far from the Evidence? The EUPHAS 2 Project

Erica L. Martin[a] · Dinna N. Cruz[b,c] · Gianpaola Monti[d] · Gianpaolo Casella[d] · Sergio Vesconi[d] · V. Marco Ranieri[a] · Claudio Ronco[b,c] · Massimo Antonelli[e]

[a]Department of Anesthesiology and Critical Care, University of Turin, Ospedale S. Giovanni Battista-Molinette, Turin, [b]Department of Nephrology, Dialysis and Transplantation, San Bortolo Hospital, [c]International Renal Research Institute (IRRIV), Vicenza, [d]Department of Intensive Care 'G. Bozza', Ospedale Niguarda Ca' Granda, Milan, and [e]Department of Intensive Care and Anesthesiology, Catholic University of Sacred Heart, Rome, Italy

Abstract

Since 1994, a polystyrene fiber cartridge used for extracorporeal hemoperfusion, to which polymyxin B is bound and immobilized, has been used in septic patients in order to absorb and remove circulating lipopolysaccharide, thereby neutralizing the effects of this endotoxin. This therapy gradually gained acceptance as the amount of evidence increased from initial small clinical studies to a carefully conducted systematic review, and ultimately to the multicentered randomized clinical trial conducted in Italy, entitled the EUPHAS Study (Early Use of Polymyxin B Hemoperfusion in Abdominal Septic Shock). While the conclusions of this initial randomized controlled trial were in agreement with previous studies, it possessed some important limitations, including a slow accrual rate, enrolling only 64 patients between 2004 and 2007, inability to blind treating physicians, and a premature study termination based on the results of the scheduled interim analysis. These limitations resulted in a modest patient sample size, which may have overestimated the true magnitude of the clinical effect. Apart from Japan, Italy is the current primary user of polymyxin B-hemoperfusion in the treatment of sepsis, with about 600 cartridges being used per year. However, no structured collection of data has been attempted, resulting in the an opportunity to understand the effects of polymyxin B-hemoperfusion on a large, diverse sample size. In response, Italian investigators and users of this treatment have designed a new prospective multicentered, collaborative data collection study, entitled EUPHAS 2. The aim of the EUPHAS 2 project is to collect a large database regarding polymyxin B-hemoperfusion treatments in order to better evaluate the efficacy and biological significance of endotoxin removal in clinical practice. Additionally, this study aims to verify the reproducibility of the data currently available in the literature, evaluate the patient population chosen for treatment and identify subpopulations of patients who may benefit from this treatment more than others.

Endotoxemia is defined as the presence of endotoxins, or toxic structural components of certain bacteria, in the blood, which can result in organ dysfunction, shock and death. Endotoxemia can be caused directly by a Gram-negative infection or indirectly through impaired intestinal epithelial barrier function, and remains extremely common in critically ill and injured patients [1]. Lipopolysaccharide (LPS) is the most common endotoxin, which in healthy subjects is scarcely detectable in the blood [2]. During sepsis, however, it can increase as much as 1,000-fold, even in the absence of an identified Gram-negative infection [3]. Furthermore, these high levels of LPS are associated with worsened clinical outcome [4].

Polymyxin B is an antibiotic used to restrain Gram-negative infections through its ability to bind with high affinity to the LPS of the bacterial wall, thereby increasing its permeability and inducing bacterial death. Due to it ability to strongly bind to LPS, polymyxin B was originally proposed as a treatment to remove circulating endotoxin; however, it was quickly discovered that the systemic use of polymyxin B was highly neurotoxic and nephrotoxic. Therefore, pharmaceutical companies have attempted in recent years to design new molecules that can mimic the affinity of polymyxin B to LPS, while avoiding harmful side effects. Unfortunately, many of these antiendotoxin therapies, including monoclonal antibodies or LPS-neutralizing proteins, have failed to demonstrate a significant biological benefit in clinical trials [5].

The first successful exploitation of the unique properties of polymyxin B was by Toray Medical Ltd., a Japanese company leader in the extracorporeal dialysis filters. Toray Medical bound and immobilized polymyxin B to a polystyrene fiber cartridge used for extracorporeal hemoperfusion, during which the fixed polymyxin B absorbs and removes circulating LPS, thereby neutralizing the endotoxin. This treatment has been used in Japan since 1994 and has gradually gained confidence throughout Europe and North America. This acceptance has grown as the amount of evidence increases, which began with several small clinical studies and has led to an important systematic review, published in 2007 [6].

The EUPHAS Study

These efforts led to a multicenter randomized clinical trial, entitled the EUPHAS study (Early Use of Polymyxin B Hemoperfusion in Abdominal Septic Shock), which was published in June 2009 in JAMA (the Journal of the American

Medical Association) [7]. This study was conducted in 10 Italian intensive care units (ICUs) and is the largest multicenter randomized controlled trial using direct hemoperfusion with polymyxin B on patients with abdominal septic shock. Cruz et al. [7] focused their inclusion criteria to study a very homogenous and severely ill patient population, who were likely to have high endotoxin levels from a definitive surgical source.

The main results showed that mean arterial pressure increased (76 to 84 mm Hg; p = 0.001) while vasopressor requirement decreased (inotropic score, 29.9 to 6.8; p = 0.001) over the 72-hour study period in the polymyxin B group, but not in the conventional therapy group (mean arterial pressure: 74 to 77 mm Hg; p = 0.37, inotropic score: 28.6 to 22.4; p = 0.14). The PaO_2/FiO_2 ratio slightly increased (235 to 264; p = 0.049) in the polymyxin B group, but not in the conventional therapy group (217 to 228; p = 0.79). Additionally, SOFA scores improved in the polymyxin group, but not in the conventional therapy group (change in SOFA: –3.4 vs. –0.1; p = 0.001). The 28-day mortality rate was found to be 32% (11/34 patients) in the PMX group compared to 53% (16/30 patients) in the conventional therapy group (unadjusted HR: 0.43, 95% CI: 0.20–0.94; adjusted HR: 0.36, 95% CI: 0.16–0.80).

In conclusion, the EUPHAS study results are in agreement with those shown in the systematic review of Cruz et al. [6] and further confirm an effect of rapid improvement in overall organ function, particularly within the cardiovascular system. Nonetheless, the study had some important limitations (as described by the authors), which must be taken into consideration. First of all, although it is also one of the strengths of this study, the highly targeted patient population led to a slow accrual, enrolling only 64 patients between 2004 and 2007. Secondly, due to the nature of intervention, it was not possible to blind treating physicians to the allocation group of the patients, although data analysts remained blinded to experimental groups. Finally, this study was stopped prematurely based on the results of the scheduled interim analysis, which met the accepted standard criteria for early termination. These limitations resulted in a modest patient sample size, which may overestimate the true magnitude of the clinical effect. Moreover, in his editorial, John Kellum [8] stated that the EUPHAS study was not designed as a definitive trial with a centered patient-point, but instead was to determine whether polymyxin B hemoperfusion would result in improved mean arterial pressure and less requirements for vasopressors in patients with septic shock presumed to be from abdominal infection.

The EUPHAS 2 Study

Since its arrival on the market in 1994, when the Japanese health insurance system first approved polymyxin B direct hemoperfusion (PMX-DHP), approximately 75,000 cartridges have been used. Combining both clinical trials and the various surveillance systems of medical devices, no noteworthy side effects or adverse events have ever been reported. Nevertheless, the international scientific community remains skeptical regarding the effectiveness of extracorporeal endotoxin removal therapy in clinical practice.

Apart from Japan, Italy is the primary country currently using polymyxin B hemoperfusion in the treatment of sepsis, with about 600 cartridges used per year (data from Estor Spa and Toray Medical Ltd.). However, to date, no structured collection of data has been attempted, resulting in a lost opportunity to understand the effects of PMX-DHP on a large diverse sample size. In response, the most established Italian investigators and users of polymyxin B hemoperfusion have decided to design a new prospective multicenter, open, collaborative data collection study, entitled EUPHAS 2.

EUPHAS 2 will be coordinated by a scientific steering committee (SSC) composed of four members: two researchers of the original EUPHAS trial and two other experienced PMX-DHP users who were not involved in the previous EUPHAS trial.

The project will be a collaborative web database, where every registered user will be allowed to introduce and analyze data of patients treated in their center. A statistical overview of the whole database will be automatically available and updated daily on the web. Data from the entire database will only be available to the SSC; however, the data will be presented to all registered users annually at an organized EUPHAS 2 user meeting. This meeting will provide the opportunity to decide how to use and disseminate the collected PMX-DHP data.

The aim of the EUPHAS 2 project is to collect a large quantity of data regarding PMX-DHP treatments in order to better evaluate the efficacy and biological significance of endotoxin removal in clinical practice. Additionally, this study aims to verify the reproducibility of the data currently available in literature, evaluate the patient population chosen for treatment and identify subpopulations of patients who may benefit from this treatment more than others.

In order to address these objectives, all patients with severe sepsis or septic shock of any origin, dysfunction of one or more organs, and high levels of endotoxin activity, as determined using the Endotoxin Activity Assay (EAA™) with an EAA™ value of greater than 0.6 [4, 9]. This will allow for the analysis of possible correlations between the type/location of initiating infection and/or time of treatment to the effectiveness of PMX-DHP therapy.

The EUPHAS 2 Study will be characterized by two phases. Phase 1 will be a retrospective collection of data from severe sepsis or septic shock patients treated with polymyxin B hemoperfusion in Italian ICUs within the last 3 years. The aim of this phase is to obtain historical data for at least 250 patients. This preliminary data collection is estimated to require 6 months to complete, after which the SSC will analyze the collected data in order to define the criteria conditions which will be utilized in the subsequent prospective data collection (phase 2).

Phase 2 will be a prospective, multicenter, open, web database and data collection of all patients treated with PMX-DHP in participating Italian and European ICUs. All patients will be completely managed according to standard protocols of each ICU, which includes fluid infusion, vasopressor administration, mechanical ventilation, antimicrobial chemotherapy and support of renal function, such as artificial CRRT and/or hemodialysis. Therefore, the inclusion of patients in the database will not, under any circumstances, impose any change on patient management, and each patient will receive a follow-up for 29 days (from day 0 to day 28) following entry into the study.

The database structure will be created to easily collect data using a specific case report form that will include:

- Inclusion data: demographics, date of diagnosis of septic shock and endotoxin activity value, results of biological cultures, underlying diseases, main treatments and concomitant treatments with other medical devices, and severity scores of the patient.
- Pretreatment, post-treatment and follow-up data: vital signs, vasoactive pharmaceuticals, diuresis, SOFA score, hemodynamic variables, hemogas analysis, adverse events, concomitant care (MV,CRRT, antibiotics), ICU length of stay, and outcome.

Since in Italy alone, over 600 polymyxin-B based cartridges are used each year, with patients normally receiving two PMX-DHP treatments, we have modestly estimated that the database could collect one third of these treatments, indicating a sample size of approximately 100 patients annually.

Database Statistical Analysis

Some information will be continuously processed and available in 'real-time' for all users of the database. This will include the number of registered centers, the number of patients traced and statistics on the type of patients enrolled by severity, age, gender and diagnosis. Individual centers will maintain ownership and management of all data entered by that unit, i.e. each center can visualize, analyze or modify their personal data at any time.

The collected raw data will be analyzed each year by the SSC in order to monitor the use of PMX-DHP therapy in clinical practice, with particular reference to the variation of effectiveness of the treatment with respect to:

- Time of intervention
- Patient severity at the time of the first PMX-DHP treatment
- Primitive pathology
- Assessment of endotoxin activity with EAA™.

The efficacy of the treatment will be assessed using the same criteria as the original EUPHAS study, namely:

- A 20% decrease of the vasopressor dependency index at 72 h

- A reduction of 3.5 points of delta SOFA score at 72 h
- Renal function: significant improvement of urine output and/or frequency of CRRT
- Respiratory function: significant improvement in oxygen metabolism, blood gases and/or mechanical ventilation-free days
- Improvement in the ICU length of stay and/or ICU-free days (indicating the actual need for days of intensive care)
- Patient prognosis.

This analyzed data will be presented to all project participants during the annual user group meeting, and upon agreement by the SSC, will be submitted for publication.

The EUPHAS 2 project web-site and database will be created in both Italian and English, allowing the collection of data not only from Italian users but also from users of various countries as PMX-DHP use expands across Europe. New sites will be added to the study upon approval of the SSC.

Conclusions

Although polymyxin B hemoperfusion treatment of septic patients has shown early therapeutic promise, larger multicenter studies are still required to confirm the encouraging results of the EUPHAS study in other patient populations [6]. The new EUPHAS 2 study, involving prospective data collection, will provide more evidence regarding the use of DHP-PMX since it will gather a large quantity of data from many different centers and from a wide population range of septic shock patients, involving any patient with high endotoxin activity. It is therefore our aim that the EUPHAS 2 study will be able to provide a definitive statement regarding the use of polymyxin B hemoperfusion for the treatment of endotoxic septic shock, provide support for the FDA randomized controlled trial proposed to start in the USA, and ultimately determine the role of PMX-DHP in septic therapy.

References

1 Kellum JA: A targeted extracorporeal therapy for endotoxemia: the time has come. Crit Care 2007;11:137.

2 Klein DJ, Derzko A, Seeley A, et al: Marker or mediator? Changes in endotoxin activity as a predictor of adverse outcomes in critical illness. Crit Care 2005;9(Suppl 1):161.

3 Opal SM, Scannon PJ, Vincent JL, et al: Relationship between plasma levels of lipopolysaccharide (LPS) and LPS-binding protein in patients with severe sepsis and septic shock. J Infect Dis 1999;180:1584–1589.

4 Marshall JC, Foster D, Vincent JL, et al: Diagnostic and prognostic implications of endotoxemia in critical illness: results of the MEDIC study. J Infect Dis 2004;190:527–534.

5 Manocha S, Feinstein D, Kumar A: Novel therapies for sepsis: antiendotoxin therapies. Expert Opin Investig Drugs 2002,11:1795–1812.
6 Cruz DN, Perazella MA, Bellomo R, et al: Effectiveness of polymyxin B-immobilized fiber column in sepsis: a systematic review. Crit Care 2007;11:R47.
7 Cruz DN, Antonelli M, Fumagalli R, et al: Early use of polymyxin B hemoperfusion in abdominal septic shock: the EUPHAS randomized controlled trial. JAMA 2009;301: 2445–2452.
8 Kellum JA, Uchino S: International differences in the treatment of sepsis: are they justified? JAMA 2009;301:2496–2497.
9 Valenza F, Fagnani L, Coppola S, et al: Prevalence of endotoxemia after surgery and its association with ICU length of stay. Crit Care 2009;13:R102.

Erica L. Martin
Dipartimento di Anestesiologia e Rianimazione
Università di Torino, Ospedale S. Giovanni Battista-Molinette
Corso A.M. Dogliotti 14
IT–10126 Torino (Italy)
Tel. +39 011 633 4005, Fax +39 011 696 0448, E-Mail ericaleanne.martin@unito.it

Author Index

Subject Index